EXPERIMENTING
IN
PSYCHOLOGY

EXPERIMENTING IN PSYCHOLOGY

Robert Gottsdanker

University of California
Santa Barbara

Prentice-Hall, Inc.
Englewood Cliffs, New Jersey 07632

Library of Congress Cataloging in Publication Data

GOTTSDANKER, ROBERT M
    Experimenting in psychology.

    Bibliography: p.
    Includes index.
    1. Psychology—Experiments. 2. Experimental design.
3. Psychometrics. I. Title.
BF181.G67   150'.7'24   77-26229
ISBN 0-13-295501-6

PRINTED IN THE UNITED STATES OF AMERICA

10 9 8 7 6 5 4 3 2 1

Prentice-Hall International, Inc., *London*
Prentice-Hall of Australia Pty. Limited, *Sydney*
Prentice-Hall of Canada, Ltd., *Toronto*
Prentice-Hall of India Private Limited, *New Delhi*
Prentice-Hall of Japan, Inc., *Tokyo*
Prentice-Hall of Southeast Asia Pte. Ltd., *Singapore*
Whitehall Books Limited, *Wellington, New Zealand*

To my wife, Josephine,
and to others
whose influence has been great:
my parents and my children,
my teachers and my students

CONTENTS IN BRIEF

CONTENTS IN DETAIL

PREFACE

This is a book on *doing* experiments in psychology. It is not meant to "cover" particular subject-matter fields, such as sensory processes, or learning, or social behavior. Rather, it is concerned with the *methods* of experimentation that cut across all of the subject-matter fields. So it is a book with a methodological emphasis.

Those of you familiar with other books having a similar emphasis will notice a difference at once. This book does much less talking about SCIENCE. Yet you will find that it is organized much more like the textbooks in the longer-established sciences. It starts with easy experimental problems, having simple solutions. Step by step, it goes on to increasingly challenging and rewarding experimental problems. Each step is built on the foundation of the previous steps.

Although the direction is toward increasingly generalizable experiments, applied experimentation is never downgraded. It occupies almost half of the book. The author firmly believes that a good applied experiment is worth a dozen whose main virtue is that they are not applicable. Certainly, with the new emphasis on evaluation of programs, applied research is coming into its own.

There are some special advantages of a building-block approach for a methodological course in experimental psychology. The other alternative has been to treat various "topics," such as *measures of behavior, subjects, experimental designs, and treatment of data* in separate chapters. When that is done, there can be no comprehensive knowledge about any kind of experiment until the last chapter is covered. The student is required to wait and wait and to remember and remember until the parts can be put together, in order to understand fully even the simplest experiment. Here, from the start, whole experiments are described at a level that is within the students' grasp. There is never any need to wait for a missing part of knowledge. The advantage is especially great when there is a laboratory portion of the course. Close coordination with text and lecture may be attained by a graded series of experiments. In fact, worthwhile laboratory experiments may be coordinated with the first chapter.

The central theme of the book is a synthesis of the ideas of *validity* and of the *perfect experiment*. Campbell and Stanley's (1962) distinction between *internal* and *external* validity is systematized, using rather stricter definitions than have hitherto been published. It is hoped that something more has been accomplished than merely adding new confusion to the sufficient amount already present on this matter. Keppell's (1968, p. 23) wry observation on the "ideal experiment" has been expanded to include other types of perfect experiments. In brief, an experiment is said to possess one of the two kinds of validity as it approaches one of the several types of perfect experiment.

The inclusion of statistical supplements to the chapters acknowledges the importance of data analysis and statistical inference in experimentation. They have been written to accommodate the needs of a variety of students and the inclinations of a variety of instructors. They may serve as refreshers for students who have already taken a course in statistics. For such students and also for those who are taking a statistics course concomitantly with the experimental course, they should help integrate the two areas. For students who will take statistics at a future time, the supplements should be useful as a temporary substitute for the fully developed course. Those instructors who do not wish to combine statistics and experimentation should not feel obliged to use the supplements at all. The book does not depend on the supplements. One word of advice is not to look for the statistical supplement to Chapter 5. There isn't any!

My colleagues at Santa Barbara have all been helpful to me as I have "bounced ideas off them." However, Professor Marilynn Brewer deserves a special word of gratitude. In our numerous discussions on statistical matters she has excelled in erudition, limpidity, and patience. To Professor James J. Jenkins, Series Advisor to Prentice-Hall, I give my heartfelt thanks for his unswerving dedication to fundamentals; I also thank the following reviewers for their helpful suggestions: Professor Charles E. Cliett, East Carolina University; Professor Edward Domber, Drew University; Professor Robert W. Fernie, Phoenix College; Professor Joseph B. Hellige, University of Southern California; Professor Mark Sobell, Vanderbilt University. Finally, I am endebted to Mr. Theodore K. Jursek, Psychology Editor for Prentice-Hall, for the confidence he has shown in me.

I am grateful to the Literary Executor of the late Sir Ronald A. Fisher, F.R.S., to Dr. Frank Yates, F.R.S., and to Longman Group Ltd., London, for permission to reprint Table IV from their book, *Statistical Tables for Biological, Agricultural and Medical Research* (6th edition, 1974).

R. G.

EXPERIMENTS THAT DUPLICATE THE REAL WORLD

A power loom for weaving cotton cloth is a noisy machine. Put yourself in a weaving shed in Lancashire, England. Packed closely together are long rows of looms. "The noise level in a typical textile mill is above the level at which deafness occurs following prolonged exposure" ("Textile Industry," 1974). In other words, the sound is deafening. What is the effect of this din on the behavior of a weaver? Even looked at coldly—in terms of production—doesn't it seem reasonable that reducing the noise would be helpful? That is the background for the first experiment that we will discuss.

This is a book about experiments in psychology, practical ones like that on weaving and some in the laboratory tradition of "pure" research. When you have completed the book, you should be able to read typical articles on experiments with understanding. Further, you should be able to decide whether an article has made a valid point. Finally, you should be able to do many kinds of experiments yourself.

In this first chapter, you will be given the tools to do a simple kind of experiment. The best test of whether you can use the tools is for you to do an experiment and to have your efforts evaluated by someone who has already done some experiments. However, a good deal of your understanding of this chapter can be tested somewhat more easily. At the end of the chapter, you will be asked questions in three main areas:

1. Why this book starts with experiments that duplicate the real world.
2. Planned manipulation and documentation: the two keys to the experiment.
3. The basic words in the language of the experiment.

AN EXPERIMENT ON WEAVING AND NOISE

Two investigators of the Industrial Health Research Board of Great Britain, H. C. Weston and S. Adams, conducted an experiment in a Lancashire weaving shed to find the effect on production of cotton cloth of lowering the noise level for a weaver (1932). The looms are bound to be noisy as the heavy shuttles fly back and forth adding to the woven cloth. As the investigators point out (p. 40), nothing could be done to reduce the noise of the looms, but each weaver could wear plugs called Mallock-Armstrong *ear defenders*. They are devised to allow the transmission of "small sounds so that it is possible to carry on a conversation without removing them."

Without ear defenders, the sound level in the shed, as measured with an audiometer, was 96 decibels. Use of the ear defenders lowered the sound level to 87 decibels. This is really a larger reduction than it might seem: The noise sounded less than half as loud. Still, what was accomplished was to make the noise environment more like having someone continually shouting in your ear rather than like standing beside a rushing freight train. In ordinary face-to-face conversation, the sound level is about 60 decibels.

Ten weavers, seven women and three men, served as subjects for the experiment, which extended over a period of 26 weeks. The experiment started on Monday, May 27, 1929, and ended on Friday, November 29, of the same year. During the first week, ear defenders were not worn; during the second week, they were worn; and so on, for alternating weeks, without and with ear defenders.

A weaver tended four power looms. Her (most of them were women) job was largely to reshuttle and to repair broken ends of thread. The shuttle is loaded with thread that it lays down inside the angle formed between a "floor" and sloped "ceiling" made up of alternating lengthwise threads. Each cross-length of thread carried by the shuttle is called a *pick*. The movement of the shuttle back and forth across the loom is automatic. When the shuttle runs out of its load of thread, it must be stopped and another load of thread put in. This is called *reshuttling*. Sometimes a thread from the shuttle breaks, and the weaver must tie together the broken ends of thread.

In the particular operation studied, the shuttle would move across the loom 11,759 times an hour if there were never any stops; that is the number of possible picks per hour. The efficiency of the weaver in reshuttling and tying ends is shown by how *few* picks she misses per hour. By and large, the weavers were very efficient, tending to miss only about 1,000 picks per hour, or fewer than 10 percent. The number of actual picks was measured by a counter that was advanced one step each time the shuttle reached either end of the loom. There was such a counter on one of the four looms tended by a weaver.

Generally speaking, performance was somewhat higher in the 13 weeks in which ear defenders were worn than in the 13 weeks in which they were not. However, the amount of effect was highly individual; a few of the weavers were helped very little. Moreover, five of the weavers reported being disturbed by the noise when not wearing ear defenders, and the other five reported being largely indifferent to the noise.

In view of these differences among weavers, the experiment should be looked on as ten separate experiments, one for each weaver. In practical terms, the experiments are, in fact, individual. Each weaver is concerned with her own productivity and comfort, and the decision of whether to wear ear defenders is entirely an individual matter.

We will look at the result for the weaver referred to as Subject D. She was one of the five who was disturbed by the noise. Her comment after the end of the experiment was: "I like them very much and wish to continue wearing them. I feel better with them in use, but I am always conscious that I have them" (p. 47). Some other weavers who liked the ear defenders had even more favorable—and "colourful"—remarks. Subject C: "OK. They're good on certain days, and are restful. They put me in order when I'm feeling 'nattered'; I should like to continue with them" (p. 47). Subject K: "They make a big difference to me, and I always feel better when I am wearing them. I don't feel 'nattered' when I am wearing them. My mother has noticed that I have been better in temper. I feel that they steady the nerves (This weaver asked for a pair of ear defenders when the experiment ended, and now she wears them daily in the shed)" (p. 47). Nevertheless, Subjects C and K, despite being less "nattered" showed less improvement in production with ear defenders than did Subject D.

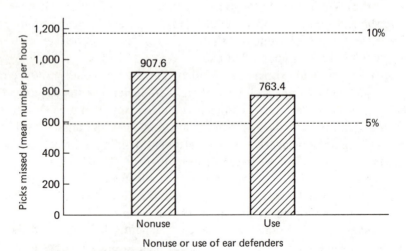

Figure 1.1 Performance by Subject D, nonuse and use of ear defenders, each on 13 alternate weeks.

The overall results on output for Subject D are shown in Figure 1.1. Over the 13 weeks with nonuse of ear defenders, she missed, on the average, 907.6 picks per hour; over the 13 weeks with use of the ear defenders, this was reduced to 763.4 picks per hour. It is also shown, in Figure 1.1, that both of these performances represent between 5 percent and 10 percent of possible picks missed per hour. Since the wearing of ear defenders reduces the chance of deafness, since Subject D is disturbed by the noise without ear defenders, and since she was somewhat more efficient wearing the ear defenders, there is little question that she should wear them in the future. The only negative aspect is that she is always conscious that she is wearing them.

FROM EVERYDAY LIFE TO EXPERIMENTS

As in any experiment, in the one just described a limited amount of evidence was collected to reach conclusions that hold more generally. Thus, we say that an experiment is done in order to *generalize*. For this particular weaver, the conclusion was that it will be profitable to wear ear defenders in future work. In this case, the results are applied to that weaver's future world of work, after the experiment is all over.

The experiment is much like the way we reach conclusions in everyday life. We are all guided in our future actions and attitudes by limited experiences. If I encounter on more than one occasion a person who smiles warmly at me, shakes my hand, and seems concerned about my well-being, I judge that person to be friendly and am apt to seek him out whenever I see him, whether at a party, at a conference, or just walking along the street. Of course, one may draw the wrong conclusion. The friendly person might really only want to sell me life insurance. Yet, we would not know what to do in any new situation if we did not generalize what we have gained from previous ones. Most of us have learned that saying "Hello" generally results in further conversation; saying "Don't you look odd!" tends to shut off conversation.

Our first example of an experiment, as well as the two other experiments that will be described in this chapter, are close to everyday life. In fact, they are duplicates of the *real world* for which conclusions are drawn. Nothing was changed about the scene or the work in the weaving shed in order to do the experiment. After the experiment was over, there was again no change. We would have seen the same weaver, at the same loom, weaving the same kind of cloth, over the same work week. The experiment duplicated two real worlds of the future for each weaver: one with ear defenders, one without ear defenders. The generalization provided was minimal. Results could not be applied to other persons or other kinds of work. In most experiments, the effort is made to obtain results that can be generalized much further. The experimental approach is useful to the extent that generalization is justified. That, in fact, is a major theme of this book.

Yet an experiment that duplicates the real world is a good one to use for starting to learn about experiments. The first reason is that it is the kind of experiment that is closest to the world of application. For this reason, you will not have to ask whether the subject was a good representative of the individuals to whom the results will be applied; the results of a given subject are applied only to the same person. By putting this question off until later, you will be able to concentrate your attention on the basic ideas that hold for all experiments.

The second reason for starting with this kind of experiment is its nontheoretical nature. Such experiments are done for very practical purposes. The efficiency and comfort of the individual weaver was the concern of this experiment. The other experiments to be described in the chapter are done to help one person find a better way of memorizing piano pieces and to help another person find the best-tasting brand of tomato juice. These experiments, then, are not devised to deal with more abstract questions of theory. Thus, you will be able to learn about experimentation without at the same time having to learn about some psychological theories.

The third reason for starting with experiments that duplicate the real world is to get you started at once in doing experiments. You can select your experimental problem by thinking of how your way of doing things might be improved. You can find your subject by looking in the mirror!

PLANNED MANIPULATION, THE FIRST KEY TO THE EXPERIMENT

Even though an experiment may duplicate everyday life in what the subject does and in what surroundings, it must be more systematic in other ways. The weavers' experiment illustrates the first essential of an experiment, *planned manipulation*. The weeks in which ear defenders were worn were decided on ahead of time. Consider the improvement in the evidence obtained over that which would be available if there had been no advance plan for deciding when to wear ear defenders.

Suppose it were left to the weaver to try out ear defenders from time to time to find whether she benefited. She might wear them when she wanted to do especially well, perhaps because she had worked poorly the day before. If production turned out to be high at those times, the effect could be due only to her motivation. It would also be possible that she would decide to use ear defenders at times when she felt especially "nattered" and wanted less noise. If production turned out to be low at those times, the effect might be due only to her irritability.

Without planned manipulation, then, all that could be known is that there is some sort of *correlation* between output and use or nonuse of ear defenders. In one case, high production with ear defenders was a result of the state of another variable, motivation of the weaver. In the second, low production with use of ear defenders was a result of the state of a different variable, irritability of the weaver. In the first case, production was *positively* correlated with the use of ear defenders; in the second case, it was *negatively* correlated. Clearly, the mere existence of a correlation is poor evidence that a given variable has an effect on behavior. In contrast, through planned manipulation, in which it was decided in advance that ear defenders would be worn on alternate weeks, there was convincing evidence of the effect of the wearing of ear defenders on production.

THE BASIC WORDS IN THE LANGUAGE OF THE EXPERIMENT

We have already used the word *variable* several times but have not defined it. We shall remedy that at once. It is one of the main words in the language of the experiment. The purpose of this language is to help us think and talk about an experiment in general terms; that is, with words that can be applied to all experiments. Use or nonuse of ear defenders is an example of a variable. Mean number of picks missed per hour is another variable. The term *variable* means nothing more than that there are differences in some entity. Use or nonuse are clearly different in respect to the entity, wearing of ear defenders. Similarly, 700 and 900 are different in respect to the entity, mean number of picks missed per hour. Such entities, then, are called *variables*. However, the two just described are different kinds of variable.

Independent Variable

Use or nonuse of ear defenders is an *independent* variable. This is what the experimenter manipulates planfully. This particular independent variable consists of two *treatments*. Use of ear plugs is one treatment, nonuse of ear plugs is the other treatment of this independent variable. Another word used for treatment is *level*, especially when the treatments differ in some quantitative way. For example, there might have been four kinds of ear defenders, one that reduced the sound to 87 decibels, another that reduced it to 78 decibels, and two others that reduced it to 69 and to 60 decibels. If an experiment were done to compare the effect on production of these four types of ear defenders, the independent variable would be *sound level* with use of ear plugs. The four levels would be 87, 78, 69, and 60 decibels. Two other words that are used occasionally instead of treatment are *condition* and *factor*.

Dependent Variable

Mean number of picks missed per hour is a *dependent variable*. It is said to have different *values*, depending on which treatment was used. For each treatment, then, there is one value of the dependent variable. The main fact found in an experiment is the relation between the independent and dependent variables. This relation is seen most clearly in the kind of graph shown in Figure 1.1. You should note that the independent variable is placed on the horizontal axis and the dependent variable is measured off on the vertical axis. Other words for the horizontal and vertical axes are *abscissa* and *ordinate*.

You should, in your own presentation of experimental results, be exact in your definition of the independent and dependent variables. For the present experiment, this poses no problem for the independent variable, which is simply the wearing of ear defenders, with the two treatments labeled *use* or *nonuse*. The *dependent variable* is not a vague term such as *output* or *production*. Several times each day there was an *assessment*—the number of picks were counted. The dependent variable was one particular way of *combining the assessments* for each treatment, finding the mean number of picks missed per hour. A different dependent variable could have been obtained with a different way of combining the assessments. For example, it could have been the number of weeks in which there were more than 800 picks missed per hour. If that dependent variable had been used, the value with use of ear defenders would have been 5 weeks and the value with nonuse would have been 9 weeks for Subject D.

Experimental Hypotheses

Like many experiments, this one started with "an educated guess." The guess—in respect to production—was that more cloth would be woven if ear defenders were used than if they were not, by reducing the unproductive time percentage. When this guess is put into the concrete terms of an actual experiment, it is called an *experimental hypothesis*. In the present experiment, the experimental hypothesis was that on weeks in which ear defenders are worn there will be a smaller number of picks missed per hour than on the weeks in which they are not. The experimental hypothesis is tested by the relation found between the independent and dependent variables.

Whenever an experimental hypothesis is proposed there is automatically the *counterhypothesis*, that the relation is just the opposite. Here it would be that the number of picks missed will be lower with nonuse of ear defenders than with use of ear defenders. We need not, at this time, consider a third rival hypothesis that it makes no difference whether ear defenders are worn, although, as you will see in Chapter 6, this alternative hypothesis is important in more analytic or scientific experiments. Only two *rival hypotheses* need be considered for a practical decision of whether to wear ear defenders. There is really no bother or much expense in using defenders, nor if the difference in the experiment turns out small is there any important consequence of a wrong conclusion. The reasonable decision for the weaver is to wear ear plugs if they are favored at all by the experimental results and not use them if nonuse turned out better.

The experiment was done in order to reach a conclusion of which of the two rival experimental hypotheses was true and which was false. Limited evidence was used to reach a conclusion in respect to the two rival hypotheses. For the weaver described, it was concluded that the hypothesis that there would be higher weekly output, as shown by a smaller unproductive time percentage, was true, and the counterhypothesis that the relation was just the reverse was false. What happened in the study was clear enough. However, you should recognize that experimental hypotheses go beyond the period of the study. The present experimental hypothesis is stated in terms of the future work of the weaver for years to come. What happened in the experiment is used to reach a conclusion for the wider context.

AN EXPERIMENT COMPARING TWO METHODS FOR MEMORIZING PIANO PIECES: HOW TO DO A DOCUMENTED EXPERIMENT

A piano student named Jack Mozart wants to be able to learn so thoroughly the pieces he plays that he will be able to play them by heart. On the average, each of the pieces Jack plays takes around 15 minutes to perform. He has always tried to memorize a piece from beginning to end, looking at the music only when his memory fails him. But recently Jack has begun to feel that there may be a better method. He wonders how it would work if he divided each piece into three or four equal parts and memorized each part separately before trying to memorize the whole piece. He then decides to compare these two ways of memorizing: the whole method and the part method. Let us follow, step by step, how Jack went about finding an answer to his question.

Planning the Experiment

Jack wants to compare the two methods to see which works better for him. He first selects four pieces he will memorize in the experiment, two to be learned by each method. They are:

1. Beethoven, Sonata No. 30.
2. Beethoven, Sonata No. 31.
3. Debussy, Suite Pour le Piano.
4. Ravel, Sonatine.

Fortunately, Jack is a systematic young man, as will be seen from the way he went about things. He is already systematic in his hours of practicing. He has learned from bitter experience that neighbors often like to sleep late in the morning and to go to bed early at night. Also he knows that he cannot practice for more than three hours a day without becoming tired or bored. It all works out in his practicing from 2 to 5 P.M. Before starting the experiment, he makes the necessary preparations. First he prepares a laboratory notebook for recording all of the information of the experiment. He chooses a bound, hardcover composition book, which is a good idea because loose pages often get lost. He writes on the label on the cover, "Jack Amadeus Mozart [his full name], Experiment on Memorizing Piano Pieces: Whole Method vs. Part Method."

Jack then opens the cover and writes at the top of the facing page, "Page 1" and, just under that, the current date. He then heads this page "Plan of Experiment." In his plan, he writes the order in which he

Page 1
May 5, 1977

Plan of Experiment
Schedule

Practice Order	Method to Use	Piece	Playing Time	
1	Whole	C. Debussy	13 min	30 sec
2	Part	D. Ravel	11 min	30 sec
3	Part	A. Beethoven (30)	19 min	30 sec
4	Whole	B. Beethoven (31)	18 min	

I am going to memorize four pieces, one at a time. Practicing will always be done from 2 to 5 p.m., Monday through Saturday. I will start on a Monday and continue until I have learned all four by heart. That will be two renditions in a row, without error or looking at the music.

The four pieces above are all pieces I can play by sight, but which I have never tried to play without the music. By the whole method, I mean going back to the start whenever I can't remember what comes next. I will not go back to the start if I make a mechanical error such as not striking a key clearly. By the part method, I mean first dividing the piece into four sections of about equal length. I will then practice the first section until I can play it perfectly by heart, twice in a row. Then I will learn the second section in the same way, and so on. After that, I will spend a short time learning the transitions between the sections. I will then try to play the piece by memory from start to finish. If I make a mistake, I will work on the section or transition and then go back to the start.

Figure 1.2 Plan of experiment, methods of learning piano pieces.

intends to learn the pieces and the method of practice he will use on each. He also shows the time required to play the piece. He has selected pieces in pairs of approximately equal length and of similar style, with the same kinds of difficulty. The whole plan is shown in Figure 1.2. It can be seen that length and type of music are pretty well equated for the two methods. This is no accident: Jack intended to have two pairs of pieces, the pieces of each pair matched in difficulty. Moreover, Jack also describes how he is to use the two methods. So far, there is nothing about the experiment that Jack must keep in his head. He will be able to look at the notebook years later and know exactly how he had planned the experiment. Also, when a friend visits him he can show the notebook to the friend, and this person too will know all about the experiment from the written description. However, he does not dare take the notebook out of his practice room to bring to the friend's house, for fear of losing it.

Conducting the Experiment

On the first day of the experiment, a Monday, Jack makes some entries on Page 2 before starting to practice. He writes down the date, the time, the name of the piece, and the method of practice he is using. This original record is called the *protocol*. He then works on the piece the usual three hours or until he is able to play it twice in a row by heart. After he has finished practicing, he writes down the time and notes how well he has learned the piece. If he has learned the piece before the end of the session, he spends the rest of the session on finger exercises. He does not want to start to learn pieces in this experiment at different times in the session, since that might affect total time required for learning. Finally he writes down anything that might have affected his performance. A sample page from the middle of the experiment is shown in Figure 1.3. After the experiment has been completed, Jack will be able to tell whether by chance he had "bad days" more often when one rather than the other of the methods had been used.

Analyzing the Data

At the end of seven days of practice, Jack has learned all four pieces. Now his job is to assemble the information obtained (or *data*) so that he may compare the effectiveness of the two methods. He does this by preparing a "Table of Results." This is shown in Figure 1.4. The *mean* is simply the usual arithmetic average of the four times. It is quite

> Page 8
> May 17, 1977
>
> Started practice at 2:10 p.m. Had a phone call which
>   I couldn't end earlier.
> Continued on Piece #3.
> Started work on the fourth section.
> I played it through by heart once, but got mixed up
>   on the next time.
> Then I did play it correctly twice in a row, finishing
>   at 3:12 p.m.
> At this point I had to go to the bathroom. Returned
>   to practice at 3:18 p.m.
> It wasn't until 4:35 that I got the whole thing
>   together, that is, finishing the second rendition
>   in a row without error. I had kind of
>   forgotten some of the first section and also
>   had a little trouble in the transition between
>   the second and third.
> I then took out good old Czerny and worked on
>   exercises 18 and 25 until 5:16.
> Wasn't feeling too great. I seem to have
>   after-effects of that new Italian restaurant
>   I ate at last night. However, I wanted very
>   much to finish this piece and was able to
>   forget about how I felt most of the time.

Figure 1.3 Sample page of the protocol, methods of learning piano pieces.

clear that the part method was superior. On the average a piece was learned in 235 minutes by the part method and in 285 minutes by the whole method. On this basis, Jack decides to use this new method when memorizing any piece. Actually, this little experiment did not use up any of Jack's time, except perhaps for giving him a little more in the way of finger exercises than he would ordinarily have had (and needs).

DOCUMENTATION, THE SECOND KEY TO THE EXPERIMENT

Let us review Jack's documentation of his experiment and note what was included.

Page 9
May 20, 1977

Table of Results
Time Required to Memorize Pieces
by the Whole and the Part Methods

Whole Method

Piece	Playing Time (min.)	Total Practice Time (min.)
C	13.5	220
B	18.0	350
Sum	31.5	570
Mean	15.75	285

Part Method

Piece	Playing Time (min.)	Total Practice Time (min.)
D	11.5	160
A	19.5	310
Sum	31.5	470
Mean	15.50	235

Figure 1.4 Table of results, methods of learning piano pieces.

1. His experimental hypothesis: He can memorize pieces more quickly with the part method than with the whole method.
2. Descriptions of his independent variable. He wrote down exactly what his two treatments, whole and part methods, would be.
3. A statement of how he would assess the learning of a piece and how he would combine the four assessments for each method to obtain a value of the dependent variable.
4. An account of the other circumstances of the experiment: the pieces to be practiced, the time of day this was to be done, the days of the week, etc.
5. The schedule for the different treatments.
6. Techniques he will use in analyzing his data in order to test the

experimental hypothesis and a statement of how possible results
will be interpreted.
7. Perhaps of greatest importance, the original record of behavior,
the protocol on dated pages.
8. The analysis of results in respect to the research hypothesis.

Suppose that Jack had not been so systematic; let us see how he
might have proceeded. First, he might have worked out the plan in his
head and relied on his memory to carry it out. Second, he might not have
prepared recording forms in a permanent book but just jotted down the
data on any loose piece of paper available (perhaps not even immediately). There are so many ways to be unsystematic that we could go on and
on. However, the point has been made. A thoroughly documented experiment represents a vast improvement over a poorly documented one.

In conducting the experiment, Jack could well have forgotten exactly how he had defined the whole and part methods if they had not been
written into his plan. For some pieces he might, for example, divide them
into only two parts. Thus, the independent variable, method of practice,
would mean different things for different pieces. If Jack further relied on
his memory of just how long he had spent in memorizing a given piece he
would, at best, lay himself open to errors that might be either favorable or
unfavorable to the method used. At worst, he could have been prey to a
bias he has for or against the method.

An experiment cannot exist in somebody's head; it exists in the doc-
uments prepared for the experiment. After conducting an experiment, the
investigator should have one or more bound books that contain everything there is to say about the experiment.

AN EXPERIMENT ON BRAND PREFERENCE FOR
TOMATO JUICE: HOW TO WRITE UP AN EXPERIMENT

An example may now be given of quite a different experiment a person
may plan for oneself. Jack's girlfriend, Yoko Oyess, is not serious about
music but is very discriminating about what she eats and drinks. Over the
years she has developed the ritual of, first thing in the morning, taking a
5-½ oz. can of tomato juice out of her refrigerator, shaking the can thoroughly, pouring the juice into a special glass that she also keeps in the
refrigerator, and drinking it slowly and appreciatively while gazing out of
her window, which overlooks a rose garden.

Not all tomato juice meets her high standards. She has finally decided that the very best-tasting brand of tomato juice is J. J. Rittenhouse, which she buys in a little specialty shop for the exorbitant price of 91 cents for a six-pack. However, one afternoon in Jack's apartment she poured herself a glass of tomato juice from the refrigerator that she thought was even better than good old J. J. To her surprise, it came from the Buddy 'n' Bill Supermarket, under their own label, and the six-pack price was only 64 cents. The money was not really important; this is about Yoko's only extravagance. On the other hand, she did want to buy the best-tasting juice. Maybe the excellent taste of Buddy 'n' Bill's was a rare exception; maybe it was the temperature of the juice, maybe she was just in especially good spirits that afternoon.

She decides to do a little library research and finds that within the last year there had been two comparisons of brands of tomato juice reported in two consumer magazines. In the article in *Consumers' Concerns*, J. J. Rittenhouse was put in the top class, Best Buy, while Buddy 'n' Bill's was in the fourth class, Instant Reject. However, in the other magazine, *You, the Consumer*, Buddy 'n' Bill's gets the top rating, four stars, while J. J. Rittenhouse gets only three stars.

Clearly an experiment is in order. Yoko is almost sure that she would be swayed by the label on the can in a taste test. So she gets Jack to help her by removing the labels from cans and simply writing down numbers. He keeps the code of which number goes with which brand in his own laboratory notebook. She writes out the plan for the experiment and keeps records in the same way that Jack did. What follows is the experimental report she wrote after the experiment had been conducted.

Evaluation of Two Brands of Tomato Juice

Yoko Oyess

Sausalito, California

Abstract

An experiment was performed, with the writer as the subject,
to find which of two brands of tomato juice she evaluates
more highly, over a period of 36 days. Eighteen cans of each
of the two brands tested, J. J. Rittenhouse and Buddy 'n'
Bill's, were drunk in random order and evaluated--with the
labels removed. The mean rating was considerably higher for
Buddy 'n' Bill's. It was suggested that discrepancies in
previous taste comparisons were due to different conceptions
of the "true" taste of tomato juice.

Evaluation of Two Brands of Tomato Juice

It has been the writer's belief for at least two years
that the best-tasting brand of tomato juice on the market was
J. J. Rittenhouse. However, she recently tasted a brand,
Buddy 'n' Bill's, which seemed even more flavorful. Of
course it is difficult to be sure that this brand would be
better on the basis of a single experience. First, there
are doubtless variations in taste among cans of the same
brand. Second, beside the brand difference the single
experience occurred at a different time of day than is habitual
for the writer and the juice was possibly at a different
temperature. Still she had to consider the possibility that
a "halo effect" was operating for her longtime loyalty to
J. J. Rittenhouse, based in part on the prestigious label.

It is of interest to note that in previous taste evaluations there have been contradictory results. While Jenkins
(1975) gives a higher rating to J. J. Rittenhouse, Hall
(1976) finds Buddy 'n' Bill's to be superior. In addition to
finding the writer's own preference, a second purpose of the
experiment was to cast some light on the previous discrepancy.
Needless to say, the method of "blind" tasting was used, with
the juice poured from unlabeled cans.

Method

Subject

One subject was used (the writer), a woman 22 years of age.

<u>Materials</u>

　　　Three six-packs of 5-1/2 oz. cans of tomato juice of the
two following brands:　J. J. Rittenhouse and Buddy 'n' Bill's.
They were purchased the same day and taken from the shelves
of the market.　Each can was refrigerated for at least a day
before being tasted.

<u>Equipment</u>

　　　Refrigerator with cold control adjusted so that the reading
on a thermometer remained at a constant 37° F.　The same 8-oz.
straight-side, clear glass tumbler that had been put in the
refrigerator clean the previous evening was used.

<u>Procedure</u>

　　　<u>Experimental task</u>.　Each morning, between 7:15 and 7:30,
the subject (after brushing her teeth and rinsing her mouth
thoroughly) took a can of tomato juice from her refrigerator,
shook it thoroughly, and poured the juice into a glass that had
been placed in the refrigerator the previous night.　She then
drank the juice slowly over a period of about 45 sec.　At the
end of the time she made a rating on a 5-point scale as: (1)
Definitely poor--watery or with an "off" taste; (2) Fair--
nothing really wrong but not flavorful; (3) Good--a nice taste,
but not unusually flavorful; (4) Excellent--quite flavorful,
just about how tomato juice should taste; (5) Extraordinary--
the high point of the tomato juice experience (Callahan, Note 1).

　　　In addition, the subject wrote some free comments on the
nature of the flavor; for example, noting whether it could be

Evaluation of

4

described as fruity, or sweet, or spicy, etc.

Obtaining the sequence of trials. Prior to the series of
tasting trials, a random order for the two brands was worked
out. The initials JJ (for J. J. Rittenhouse) were written on
18 of the slips and BB (for Buddy 'n' Bill's) on the other 18.
The slips were placed in a paper bag, which was shaken vigorously
for one minute. With the subject out of the room, another
person picked out the slips one at a time. For the first slip
drawn, he wrote the number "1" with a grease pencil on the top
of the can of the brand indicated; for the second can he wrote
the number "2" on the brand indicated. The procedure was
continued until all 36 cans had been identified. He then made
a list of the sequence of brands, which was not shown to the
subject until after the experiment.

This person then removed the labels from all of the cans
by soaking them for about an hour in cool water. Any remaining
bits of the label were scrubbed off. It was fortunate that
the cans were identical in manufacture being the standard
5-1/2 oz. juice cans made by Intercontinental Container Cartel.
However, since different codes had been embossed on the bottom
of the two brands it was necessary to glue small disks of cardboard to the bottoms.

Conducting the tasting sessions. The experiment was
conducted on 36 consecutive days starting on July 1, 1977. At
7:15 a.m. of the previous day, the subject placed at the inner
right-hand corner of the bottom door-shelf of her refrigerator

can no. 1 and next to it the drinking glass. The next morning,
again at 7:15 a.m. she removed can no. 1 and the drinking glass,
and put can no. 2 in the same place in the refrigerator. She
then shook can no. 1, poured the juice into the glass, and drank
it over a period of about 45 sec. while standing and looking
through her kitchen window. She then entered her rating for
that can of juice in a laboratory notebook and wrote out a
short comment on the nature of the taste. That evening she
put the cleaned glass in its place in the refrigerator beside
can no. 2.

The procedure was continued for the 36 days of the
experiment. At no time did the subject attempt to look for
differences in appearance among the cans. There was some
variation in the time at which the subject started tasting the
juice, from 7:15 to 7:30 in the morning.

<u>Results</u>

Over all, the mean rating for Buddy 'n' Bill's brand was
higher than that for J. J. Rittenhouse, 3.6 as compared with
3.2. Examination of the two distributions of ratings shown in
Figure 1 reveals that the most noticeable difference was in the
number of "5" ratings: six in all for Buddy 'n' Bill's and
only one for J. J. Rittenhouse.

Insert Figure 1 about here

Free descriptions were analyzed as shown in Table 1. All

Evaluation of

6

qualities that were named at least four times fell into just
four classes: Fruity, Sweet, Bitter, Spicy. Of the 36 trials,
22 fell into one of those classes: 11 for each brand. What
is of interest is that Buddy 'n' Bill's was called Fruity five
times as compared with twice for J. J. Rittenhouse. On the other
hand, J. J. Rittenhouse was called Spicy five times to just
once for Buddy 'n' Bill's. A somewhat different quality of
taste is indicated.

Insert Table 1 about here

Discussion

The writer's hypothesis, that she would find Buddy 'n'
Bill's tomato juice preferable to J. J. Rittenhouse was supported.
It should be pointed out that the ratings for the latter were
still quite good. Her previous favorable opinion was not
entirely due to knowledge of the brand.

The different classifications of quality suggest the
reason why the previous investigators had reverse ratings for
the two brands. The panel of judges used in the Jenkins study
were from the Minneapolis-St. Paul region of Minnesota and those
of the Hall study were from the Dallas-Fort Worth area of Texas.
Perhaps the traditions are somewhat different. One suggestion
is that the Texas tasters were accustomed to eating fresh
tomatoes over a longer period of the year so were particularly
appreciative of a fruity flavor. But how to explain the possible

preference for "spicy" by the Minnesota judges? A possible
suggestion is based on the industry's bad experience in using
glass jars instead of cans. Apparently tomato juice that had
been stored in a can was preferred to that which had been in
a glass jar. The canned juice was stated to have more "true"
tomato juice flavor. In other words, for people in the
Minneapolis-St. Paul area, because of years of habituation to
canned tomato juice without much eating of fresh tomatoes, the
tinny flavor is not only accepted but is regarded as a positive
quality! It could very well be that what the subject referred
to as "spicy" actually should have been "metallic." Since
identical cans are used in the two brands, any difference in
this aspect of taste must be due to how the juice is processed.

   No claim is made that Buddy 'n' Bill's brand would be
preferred by other persons. Also, a note of caution is in order
on the generality of the findings. It is certainly possible
that there is variation from year to year in brands of tomato
juice or even during portions of the canning season. The
following are the conclusions reached: The hypothesis that
Buddy 'n' Bill's brand would obtain higher ratings was supported.
There was the suggestion that J. J. Rittenhouse brand sometimes
has a metallic flavor, which is enjoyed by some persons, but
which was not liked by the subject of this experiment.

Evaluation of

8

Reference Note

1. Callahan, H. Unpublished scale used at Contra Costa
    County Fair, Summer 1976.

Evaluation of

9

References

Hall, P. Comparison of 16 brands of tomato juice. You, the
    Consumer, 1976, 12, 49-61.

Jenkins, J. Score another for Rittenhouse! Consumer's
    Concerns, 1975, 33, 182-187.

Evaluation of

10

Footnote

The writer wishes to thank Mr. Jack Amadeus Mozart for
his assistance in this phase of the experiment. Needless to
say, the experiment was possible only because of his careful
contribution.

Requests for reprints should be sent to Yoko Oyess,
139 Viejo Ave., Sausalito, California 94965.

Table 1.

Classification of Free Descriptions

(terms used at least 4 times).

	Description				Total
Brand	Fruity	Sweet	Bitter	Spicy	
J. J. Rittenhouse	2	2	2	5	11
Buddy 'n' Bill's	5	3	2	1	11
Total	7	5	4	6	22

Figure Caption

Figure 1. Distributions of ratings given to the two brands of tomato juice.

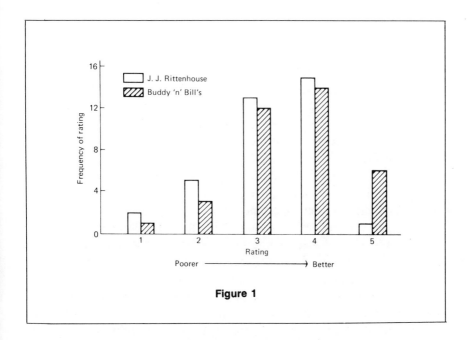

Figure 1

Note to student: On the back of all figures, the title of the article (Evaluation of Two Brands of Tomato Juice), the figure number (Figure 1), and an indication of the top of the figure would be written lightly in pencil.

THE EXPERIMENTAL REPORT: THE FINAL STEP
OF DOCUMENTATION

Even if the experiment is done only for one's own use, as in Yoko's example, the experimental report serves an important purpose. It forces the experimenter to pull all the information together and so to put the experiment in a form that may be comprehended as a whole. For the typical experiment, which has more general application, the experimental report is not only useful—it is essential. The experiment has not been completely documented until the report has been written. When it has been published, the report is called an *article*.

How often have you heard someone tell you about some amazing results found in an experiment? For example, someone might say that elephants really have very short memories. More often than not, if you try to follow up and find the article in which the result has been published, you get nowhere. Whatever you *hear* about an experiment is no more than a rumor. It is in the experimental report that the existence of the findings is documented.

Of course, it is conceivable that the experimenter would have the whole laboratory notebook published for all interested readers. However, without even taking into account the vast amount of printing that would be required, the idea is not a good one. The reason is that nobody would want to read through the many details. Further, they would have a hard time putting it together even if they were willing to read it. For the most part, an experiment is worthless unless it is communicated. The purpose of the experimental report is to communicate the experiment as effectively as possible to the people who are most likely to be the readers.

In general, this kind of communication is what the experimental report seeks to accomplish. You may now take another look at Yoko's experiment and see that it was written so that this purpose was served.

First and most important, the experimental report should be written clearly. At all times the writer should try to inform—not to impress. Second, it is divided into fairly standard sections. This way of writing experimental reports helps the writer organize the material. It also helps the reader, because he knows what to expect. With a fairly standard format, he is able to develop habits that allow him to get what he wants out of an article with greatest ease and least time spent. As you can see from the following outline, the format worked out for experiments is an effective one. If followed competently, an experienced experimenter would be able to repeat the same experiment himself. This is called *replicating* an experiment. Here, then, are the sections of an experimental report and their functions:

1. The *title* is informative of the problem studied.
2. The *abstract* gives the gist of what was done and what was found.
3. The *introductory statement* tells why the experiment was thought of and conducted.
4. The *method* section tells how the experiment was done. It should be possible to duplicate the experiment from this section, yet the description should be brief so as not to lose the more casual reader. *Method* includes:
   a. *Subjects.* Aspects that are considered important for this experiment: number, age, sex, hearing ability, etc.
   b. *Materials.* If there are different pieces to memorize, problems to solve, beverages to taste, they are described here.
   c. *Equipment.* This includes any device used for presenting conditions or for recording responses. Those descriptions relevant to the experiment are included.
   d. *Procedure.* Here are described the steps followed in carrying out the experiment, what the subject was required to do and in what surroundings, etc.
5. The *result* section presents the *analyzed data*, usually with one or more tables or graphs. The written material here usually tells the reader what to look for in the tables or graphs and why. The most important information is presented first. There may have to be some interpretation given so that the reader may make sense of the results, but emphasis should be on what was actually found.
6. The *discussion* section is where the results are interpreted and conclusions drawn. It is also the place where improvements or further experiments are suggested that might be done if the results are not completely clear. The relation of the present experiment to previous ones or to ideas that have appeared in articles or books is often commented on.
7. *Reference notes* are used for citing unpublished material, works in progress and works with limited circulation. They are listed numerically in the order of citation in the article. Within the article, they are cited by the author's last name and the word *Note* with the appropriate number.

8. *References* are listed alphabetically in standard American Psychological Association form (1976). Within the article, they are cited by the author's last name (sometimes in parentheses) and by the year of publication (always in parentheses).
9. *Footnotes* commonly used in psychological articles are usually acknowledgements and location of the author. Thus, Yoko acknowledged Jack's help. (She would also acknowledge any financial support for her study.) And she designated the address at which she could be reached for questions or for requests for reprints of the article.

SUMMARY

Three experiments were described: Use of ear defenders by a weaver, a comparison of two methods for memorizing piano pieces, and brand preference for tomato juice. They all were experiments that duplicated the real world. This kind of experiment can be seen as only a short extension of the way in which we reach conclusions from limited evidence in everyday life. We generalize from life's experiences, and we generalize from experiments.

The book starts with this kind of experiment because it is simple. Complications found in more advanced experiments do not enter the picture: for example, the question of whether the experiment is a good representation of the wider world of application and the need for understanding psychological theories. Moreover, it is the kind of experiment that the reader can do at once.

Planned manipulation was shown to be an essential element of the experiment by means of the improvement in evidence obtained in the weaver's experiment. Without it, there is only correlation between variables. On the other hand, planned treatment provides an independent variable, whose effects may be found on a dependent variable. The experimental hypothesis is tested by the relation found between the independent and dependent variables.

Documentation was shown to be an essential of the experiment by means of the improvement over a nondocumented experiment in quality of the evidence obtained in Jack Mozart's study on memorizing piano pieces. Without documentation, the experimenter must rely on his memory, which is never faultless and may even be biased. A step-by-step account was given of Jack's documentation, and then the parts were summarized.

The experimental report was treated as the final step of documentation. It was illustrated through the experiment on brand preference for tomato juice. Without a final report (called *article* when published), the experiment remains little better than a rumor. The way in which an experimental report is organized was described. The underlying idea is that of effective communication.

QUESTIONS

1. Why are experiments done that duplicate the real world?

2. Why are they the first experiments described in this book?

3. What is accomplished by planned manipulation?

4. What is accomplished by documentation?

5. Give an example showing what is meant by each of the following words in the language of the experiment.
   a. *Independent variable.*
   b. *Dependent variable.*
   c. *Experimental hypotheses.*

6. Show, by an example, how these words connect with each other.

7. What are the steps in documenting an experiment?

8. What is put into the different sections of the experimental report?

 **STATISTICAL SUPPLEMENT**

COMPUTING AND REPRESENTING THE MEANS

In an experiment on a subject's reaction time, there were 34 trials. The signal to respond was either the flashing of a light (Treatment A) or the sounding of a tone (Treatment B). The 17 trials of each of these treatments occurred equally often at random. Reaction time (RT) on each trial was measured to the nearest 1/1000 second (or millisecond, abbreviated msec).

Computing the Means

Data from the experiment are shown below. The trial numbers refer to the first, second, etc., trial with the indicated treatment.

Treatment A (Light)				Treatment B (Tone)			
Trial	RT	Trial	RT	Trial	RT	Trial	RT
1	223	10	191	1	181	10	155
2	184	11	197	2	194	11	178
3	209	12	188	3	173	12	160
4	183	13	174	4	153	13	164
5	180	14	176	5	168	14	169
6	168	15	155	6	176	15	155
7	215	16	165	7	163	16	122
8	172	17	163	8	152	17	144
9	200			9	155		

To find the mean RT for Treatment A, you simply add together the 17 RT's and divide by 17. The sum is 3,143. The mean is 3,143 divided by 17, which is 185 msec (rounded off from 184.88235). These computations can be shown by a formula:

$$M_X = \frac{\Sigma X}{N}$$

(Formula 1.1)

The assessments on the different trials are symbolized as X_1, X_2, X_3, and so on to X_{17}. Σ (capital Greek letter sigma) means to add together all of the different X's from X_1 through X_{17}. N is the number of trials or X's; here, 17. The formula is read as: "Mean X equals summation X over N." For Treatment A,

$$M_A = \frac{3143}{17} = 185$$

Representing the Means

The independent variable in the experiment was *type of signal* (light or tone). The dependent variable was mean reaction time. A bar diagram shows the relation between the independent and dependent variables. The height of a bar represents mean reaction time. (*Note*: You will compute the mean for Treatment B.)

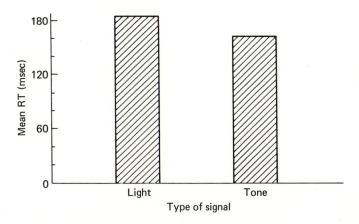

PROBLEM: Compute the mean for Treatment B.

Answer: 162 msec, rounded off.

2

BASICS OF EXPERIMENTAL DESIGN

If you wished to do an experiment to find whether having the radio on while you study will help you learn French vocabulary, you could easily do so by copying one of the experiments in the preceding chapter. Most probably you would model your experiment after Jack Mozart's. You would be careful to do your studying the same time of day, to plan your treatments in advance, and to document each of your steps. Instead of the four piano pieces, you could use four vocabulary lists in the following order: Radio, No Radio, No Radio, Radio; that is, you could use the same *experimental design* as Jack's.

Perhaps you would know the reason for some of the things you did but probably not for all, especially for the order of treatments on successive trials—the experimental design. This is hardly your fault, as no real explanations have yet been given. That lack will be made up for in this chapter. While it is possible to do an experiment by simply imitating a sample, it is far better to know what you are doing. No two experiments are identical, and copying from any model can lead to trouble. For example, Yoko could not have possibly used a regular alternating order of the two treatments (brands of juice), as was done in the experiment on the weavers (use or nonuse of ear defenders). More than likely she would catch on to which juice she was testing, just what the random order was trying to avoid. Also, unless you know the reasons for different plans or designs, you will have a hard time in deciding whether any experiment you read was done well. As you may remember, one main purpose of this book is to give you that capability.

In this chapter, we will compare the plans followed in the experiments of Chapter 1 with some poorer plans for doing these experiments. The standard of comparison will be the "perfect" experiment (which is a practical impossibility). This kind of examination will lead to the development of some general ideas to guide us in evaluating and in devising experiments. In doing this, some new words will be introduced into the language of the experiment. Finally, we will consider what is accomplished—and what is not—by the three *experimental designs* used in Chapter 1. These are the three kinds of order or sequence of the different treatments of use in the single-subject experiment.

After studying this chapter, you should be able to devise a sound experiment on your own without having to imitate somebody else's experiment. At the end of the chapter, you will be asked questions on the following matters:

1. The actual experiment as a representative of a perfect experiment.
2. The sources of internal invalidity.
3. Systematic and unsystematic effects on internal validity.
4. Methods for improving internal validity, primary controls and experimental designs.
5. Some new words in the language of the experiment.

THERE ARE PLANS AND THERE ARE BETTER PLANS

Planned manipulation is surely the first key to the experiment, but some plans are not very good. Suppose that instead of proceeding as was described, the three experiments of Chapter 1 had been done in the following planned ways:

1. If, in the experiment on the weavers, the subject had worn the ear defenders for 13 weeks and then had not worn them for the next 13 weeks.
2. If, in Yoko's experiment, she had decided to use only two cans of each juice, the whole experiment to take four days instead of 36.
3. If Jack had decided to use the part method for the first two pieces, and then the whole method for the next two pieces.
4. Or if Jack had kept the same order of treatments but had used short waltzes to memorize, rather than the longer pieces that were his real concern.

We are left with the distinct feeling that these are all poor plans as compared with the experiments as they were first described. We would be able to say exactly why the original plans were better if we had some *standard of comparison*. That need is filled by the idea of a perfect experiment. This is discussed in the next section. Then, we will see how that standard is used in evaluating our experiments.

THE PERFECT EXPERIMENT

We now have had examples of well-planned experiments and of poorly planned experiments. Is there still room for improvement in the well-planned experiment? Can the experiments be made perfect? Only one answer is needed: Any experiment can be improved, which is to say that no experiment is perfect. Experiments are *improved* as they approach perfection.

The Ideal Experiment

One way in which perfection has been defined is through the ideal experiment (Keppel, 1973, p. 23). In the ideal experiment, only the independent variable is permitted to vary (and, of course, the dependent variable, whose values are expected to change from treatment to treatment). *Everything* else is held constant, so the dependent variable can be influenced *only* by the independent variable. This certainly was not true of our three well-planned experiments. The weeks on which weavers wore ear defenders and did not wear ear defenders were different weeks, odd or even. The pieces Jack memorized by the whole and part methods were different pieces. The days on which Yoko drank the two brands of tomato juice were not the same days. In each case, something did vary beside the independent variable. In the kinds of experiments that will be described in later chapters, in which different subjects are used for each treatment, we will be able to get rid of time differences (like odd or even week) and task differences (like piece memorized). But again the conditions for the ideal experiment are not met: The subjects given each treatment are different subjects. As you will soon see, the ideal experiment is impossible. Still, it is a useful idea; it is a guide to better experiments.

For a weaver, the (impossible) ideal experiment would require her to weave cloth with and without ear defenders at the same time! Jack Mozart would memorize the same piece by the whole and part methods at the same time! In both these cases, the different values found for the dependent variable would *have* to be due to the independent variable, to the difference between treatments. Since the treatments are applied at the same time, all other circumstances, all other *potential* variables, would be held constant at the same identical level.

The Infinite Experiment

Alas, poor Yoko! For her there is a flaw even in the ideal experiment. She suspects, with good reason, that there is variation in quality among cans of tomato juice of the same brand. Even if she could do the ideal experiment by managing the trick of drinking two cans of tomato juice separately out of the same glass at the same time, her two ratings would hold only for those particular *samples* of each brand. However, she could solve the problem of variation in quality by simply performing a different impossible feat. "All" she would have to do would be to continue her experiment beyond the 36 days to an infinitely large number of days. This would take care of brand variability and would also cancel out any fluctuations in her readiness to enjoy tomato juice. This is the infinite experiment. As can readily be seen, the infinite experiment is not only impossible, it is absurd. The whole idea of an experiment is to use a *limited* amount of evidence for wider application. But, like the ideal experiment, it provides guidance.

As a matter of fact, Jack Mozart and the investigators of the weavers could also argue that they would be better off doing an infinite experiment rather than an ideal experiment. For Jack the question would remain, even if he did find the part method superior for the one particular piece in an ideal experiment, of whether this superiority would hold for other pieces. It may be that on one particular week in which an ideal experiment is conducted a weaver would prefer ear defenders and perform better with them; still, that might not hold for other weeks. But they (and you) should be warned that there is also a flaw in the infinite experiment as described: Being subjected to one treatment during the course of the experiment might very well affect the individual's behavior in the other treatment. Thus it is possible that the part method would be effective in an experiment only because of contrast with the whole method. After the experiment, only one method would be used, so contrast could not be a factor. It appears that neither the ideal experiment nor the infinite experiment is truly perfect. Fortunately, they have different flaws, so that between them they provide a standard for evaluating the very far from perfect experiments, which are the kind actually done.

The Completely Appropriate Experiment

However, the fault of doing the wrong experiment—Jack's memorizing of waltzes instead of sonatas—would not be corrected at all by either the ideal or infinite experiment. The best he could have would be a superb experiment on waltzes—and that does not make them sonatas!

To overcome this failing entirely, it would be necessary to perform the *completely appropriate* experiment, which, while not impossible, is again absurd. The pieces that Jack would memorize would be *exactly the same pieces* that he would be memorizing after the experiment. This experiment, like the infinite experiment, could serve no purpose. However, at least the argument could not be made that the pieces he learned were inappropriate.

These three types of (almost) perfect experiment are not real experiments. The ideal experiment is impossible; the completely appropriate experiment is absurd; the infinite experiment is both impossible and absurd. Their use is as "thought" experiments. They tell us what we must do to design effective experiments. The ideal and infinite experiments tell us what we must do to be more certain that the experimental results actually show the relation between the independent and dependent variables, without the intrusion of other variables. The completely appropriate experiment reminds us to check on the levels of other important variables we have held constant.

GENERALIZATION, REPRESENTATIVENESS, AND VALIDITY

As was stated in Chapter 1, in an experiment a limited amount of evidence is collected in order to reach conclusions that hold true *beyond* the experiment. That is what is meant by *generalization*. Our analysis, just completed, of perfect experiments tells us that there are at least two requirements for the conclusions to be sound. These are then the requirements for permissible generalizations to be made. First of all, there is the requirement that the relation found between the independent variable and the dependent variable be free from the contamination of other variables. Second, there is the requirement that when a potential other variable is held constant it is held so at an appropriate level, that which exists in the wider world of application.

Representativeness

We have already seen that a perfect experiment is not possible but that perfect experiments do provide guides for improving experiments. We may now ask how we are to use these guides. The answer may be found simply in *how well the actual experiment represents the perfect experiment*. First, we can examine experiments to find how well they have reduced the possibility that variables other than the independent variable are affecting the dependent variable.

In the original study of the weaver, the subject worked 13 weeks with ear defenders and 13 alternate weeks without ear defenders. In the "bad" revision, she wore ear defenders for the first 13 weeks and no ear defenders for the next 13. In the ideal experiment, she would have simultaneously worked with and without ear defenders. Certainly this ideal is approached more closely by the alternate week design. Alternation of treatments, or ABABABABAB, etc., *represents* simultaneous treatments better than does the single AB order.

In Jack Mozart's original experiment, he practiced his pieces in a whole-part-part-whole order. In the "bad" experiment, his order was whole-whole-part-part. In the original design, the *average* position for the whole and part methods were exactly the same. The positions of the whole method in the order were 1 and 4, with an average of 2.5. The positions of the part method were 2 and 3, with an average of 2.5. However, in the "bad" experiment, the positions for the whole method were 1 and 2, with an average of 1.5 and the positions for the part method were 3 and 4, with an average of 3.5. Again, the original experiment was a better *representation* of having the different treatments given simultaneously.

In Yoko's original experiment, she tasted the Rittenhouse and the Buddy 'n' Bill's brands in random order for 36 days. In the "bad" revision, she cut that down to 4 days. There is no doubt that 36 is closer to an infinitely large number than is 4. The original plan *represented* the infinite experiment better than did the revised plan.

Jack's original experiment *represented* the completely appropriate experiment better than did the "wrong" revised experiment using waltzes. While he did not use every sonatalike piece he would ever memorize, he did use that type of piece entirely; he did remain at that level. In the "wrong" experiment, using waltzes, his level of piece was a poor representation, far different from those he would learn in the completely appropriate experiment.

In summary, those experiments that were better representatives of the ideal or infinite experiments gave better information on the relation between the independent and dependent variables than those which did not. The experiment that held the level of a critical other variable closer to that of the completely appropriate experiment gave better representation of that experiment, that is, of the situation to which the experiment is to be applied.

Validity

Experiments are described as more or less valid according to how well they represent the perfect experiment. An experiment that is perfect would distinguish between true and false rival hypotheses without fail. If Jack Mozart had been able to do a perfect experiment, he would *know* for certain which hypothesis was true: *The part method is superior*; or *the whole method is superior*. Thus, the *validity* of an experiment refers to how good a job it can be expected to do in telling which rival experimental hypothesis is true and which is false.

Internal Validity   The three "bad" experiments described were lacking in *internal* validity (Campbell and Stanley, 1962). This means that they cannot be trusted to give the true picture of the relation between the independent and dependent variables. As we have seen, that happened because the effects of other variables were all too possible. An experiment that lacks internal validity thus cannot do a good job in telling which hypothesis of the relation between the independent and dependent variables is true and which is false. For example, if we cannot tell whether a weaver worked better because she wore ear defenders or rather because the weather was better, we cannot regard the experiment as giving a good basis for distinguishing between the true and false hypotheses.

The term *internal* emphasizes the basic nature of this kind of validity; another word would be *inner* validity. One can think of an experiment that lacks internal validity as being rotten to the core. If it cannot be trusted to tell us the relation between the independent and dependent variables, it is really useless.

External Validity   The "wrong" experiment Jack might have performed, using waltzes instead of sonatas, was not rotten to the core. It was a rather good experiment—on memorizing waltzes. It was not useless. Jack could use it if he decided later that he wanted to memorize waltzes with the more effective method. But the experiment was lacking in *external* validity. It did not give a good basis for distinguishing between the true and false hypotheses concerning the better method for learning sonatas.

The term *external* refers to the value of the experiment *outside* the confines of the experiment—here, to its application. In this case, the experiment was externally invalid because "sonatas" were really part of the experimental hypothesis, just as much as the independent and dependent variables were.

General Definitions   Internal and external validity will be employed as central ideas throughout this book. Their use in later chapters has been barely suggested by the distinctions that have just been made. A more formal way of defining validity and of distinguishing between the two kinds will be given to you now. Its full significance cannot be entirely clear to you until you have become familiar with a greater range of experimental problems. However, you will have a framework within which to incorporate the future usages as you come to them.

Let us start with a diagram of an experimental hypothesis:

Independent Variable . . . Relation . . . Dependent Variable
Levels of Other Variables

The hypothesis thus includes the relation and the terms on both sides of the relation. A definition of the *validity* of an experiment, including both internal and external aspects, follows. It is the extent to which the conclusion reached on the experimental hypothesis may be depended on to be that which would follow from an experiment *perfect in all respects*.

Internal validity is concerned with the relation itself, supposing there are no problems in the terms being related. The *internal* validity of an experiment is thus the extent to which the conclusion reached on the experimental hypothesis may be depended on to be that which would follow from an *ideal* or *infinite* experiment—in which the independent and dependent variables were obtained in the same way and in which there were the same levels of other variables.

However, there are also questions of *appropriateness*. The question has already been asked about the level of another variable, type of music. Later, similar questions will be asked about the independent and dependent variables. As can be seen, these have to do with the terms standing on both sides of the relation that was hypothesized. That is the concern of *external* validity. This may be defined as the extent to which the conclusion on the experimental hypothesis may be depended on to be the same as would follow from an experiment with *completely appropriate* independent and dependent variables and levels of other variables.

In the present chapter, we will be mainly concerned with the problem of internal validity. It is the problem that must be dealt with first in any experiment; there is no point in discussing external validity if internal validity does not exist. The kind of experiment introduced in Chapter 1, you may recall, was selected so that problems of external validity would be minimal. In the next chapter, experiments will be introduced that do bring to the fore questions of external validity.

No Guarantees   We can say that an experiment is valid without knowing that it has, in fact, reached the right conclusion. We can say that an experiment is invalid without knowing, in fact, that it has reached the wrong conclusion. The reason is that we cannot know in advance which rival hypothesis is true and which is false. If we did know, we would not bother doing the experiment. Jack would not have to do his experiment if he already knew which hypothesis were true: (1) part method is superior or (2) whole method is superior. In order to know whether an experiment has reached the right conclusion, it would be necessary to have this advance information.

What we have done instead is to compare *procedures* of actual experiments against those of perfect experiments. Valid experiments are better representatives of perfect experiments than are invalid experiments. Therefore a valid experiment is more *likely* to give the results that would be found in a perfect experiment. However, there will always be some risks in generalizing from limited—and imperfect—information. There will be highly valid experiments that give incorrect information on true and false hypotheses; there will be invalid experiments that give correct information. The nature of these risks and how they affect interpretation of experimental findings is discussed in several of the later chapters, especially in Chapter 6 ("Significant Results").

THREATS TO INTERNAL VALIDITY

We can now use the concept of the perfect experiment (ideal or infinite) to describe how internal validity is threatened in our real experiments. As we shall see, some of these threats cannot be avoided; they come about through procedures that must be followed in our less-than-perfect experiments. Thus, if Jack is to memorize two pieces, one of them must come first. However, there are other threats that come about only indirectly and that can be avoided if care is taken. Thus, Jack already knew better than to practice with the part method in the morning and with the whole method in the afternoon.

Variations with Time

Identified Other Variables   In the ideal experiment, the different treatments are given simultaneously. While Jack could not do that, he could at least practice the same time of day. Time of day is an easily recognized *other variable*, i.e., other than the independent variable, which might affect efficiency of practice and which can be *held constant*. If Jack were not careful, he might sometimes practice with his windows wide open on some days and not open on others. The street noise could very well affect his efficiency in practicing. This is best held constant by keeping the windows closed. In the experiment on the weavers, which went on for six months, there were identified changes in temperature and humidity. Unfortunately, the conditions of the experiment did not allow the experimenters to hold them constant. They did keep a record and did try to assess the effects of these variables. More important, their alternating experimental design did reduce the effect of this variation. Experimenters must try to identify possible other variables that could be at different levels from time to time—and do something about them. They should especially try to keep these other variables at the same constant level on each trial.

Instability over time   Try as he may, the experimenter will not be able to make one trial *exactly* like another except for having the different treatments. There is always going to be some *instability* over time. Such instability is seen in inconsistent other variables and in variations within the same treatment of the independent variable. Finally, there are always completely unidentified kinds of instability affecting the subject's response. Let us now look at some concrete examples of those three sources of instability over time.

Inconsistent other variables: Other variables that affect the dependent variable can often be recognized but cannot be directly manipulated by the experimenter. A weaver may have a "bad day" at the looms because she was up late the night before. The experimenter might try to convince her to avoid late hours until the experiment has been completed. But remember, this experiment takes six months! Jack seemed to have a bad reaction to a restaurant meal while memorizing one of his pieces; he would be more careful the following evening.

Environmental conditions will not be the same from trial to trial. The experimenters on the weavers point out:

> In cotton weaving it is well known that atmospheric conditions may affect output, since the number of yarn breakages tends to diminish as the temperature and relative humidity increase. On the other hand, beyond certain limits such atmospheric conditions, while maintaining a favorable effect on the physical properties of the yarn become physiologically unfavorable to weavers, whose working capacity may be reduced thereby to an extent which outweighs any advantage otherwise gained. [Weston and Adams, 1932, p. 56]

Hence, even with measures of temperature and humidity, the effect on output cannot be specified. The list of variables that shift in level could go on forever, including such subjective ones as feeling good or awful, during the course of an experiment. The conscientious experimenter may note some of these changes, but he cannot really do anything about them. They are out of his control. You see the forces that drive experimenters out of the real world into nice soundproof laboratories and to subjects (white rats) that he can control 24 hours a day. Yet even there, heaters sometimes fail; water bottles clog; rats come down with the "sniffles."

Introduction of an experimental atmosphere may cause continued change in behavior. That is the most general lesson for experimental psychologists to learn from the renowned "Hawthorne" experiments. Some preliminary experiments failed to establish any consistent relation between level of illumination and output in an assembly task at the Western Electric plant at Hawthorne, Illinois. A large-scale investigation of the effect of working conditions was then undertaken (Roethlisberger and Dickson, 1946). A major part of the investigation was a set of experiments in the relay assembly test toom. The task chosen "was the assembly of telephone relays, an operation performed by women, which consisted of putting together approximately 35 small parts in an 'assembly fixture' and securing them by four machine screws" (p. 20).

A special room was set up so that conditions could be controlled and behavior adequately measured. Five young women, thoroughly experienced in this kind of work, were chosen as subjects. Two independent variables were studied: rest pauses and length of working days and weeks. The women were paid according to the total number of relays assembled by the crew of five.

It was found that production continued to rise—no matter how rest pauses were arranged or how long the working days and weeks periods were—over a period of *two years*! The investigators comment about "a gradual change in social interrelations among the operators themselves, which displayed itself in the form of new group loyalties and solidarities; secondly, in a change in the relation between the operators and their supervisors. The test room authorities had taken steps to obtain the girls' cooperation and loyalty and to relieve them of anxieties and apprehensions. From this attempt to set the proper conditions for the experiment, there arose indirectly a change in human relations" (pp. 58–59).

Put in our terms, the subjects had previously operated at one level of social environment. The conditions of the experiment introduced a second level of this "other variable." This indirectly brought about continual change in the dependent variable, output, even though social environment remained constant objectively.

The independent variable: We cannot expect exact uniformity of a treatment throughout an experiment. A weaver's ear plugs will not fit as snugly on some days or even weeks as on others. Despite all attempts to make his methods of practice uniform, Jack will interpret the part method somewhat differently from piece to piece, in respect to how large the parts should be. For Yoko's experiment, each treatment was known to be variable. No two cans of the same brand of juice are identical, and sometimes the variations are large. Even in experiments in which precision is sought, there will be some variation. An electric light stimulus will vary irregularly in brightness with changes in voltage, which do occur. There may even be a consistent drift through the experiment—the light—for example—may become progressively dimmer as the bulb ages.

The dependent variable: The subject will not always make the same response to the same treatment. This will be the case even if the experimenter is unusually clever in eliminating inconsistency in other variables and is letter perfect in preventing fluctuations within a treatment.

Such instability may be demonstrated dramatically with graphs drawn for two experiments. In Figure 2.1 are shown the weekly outputs for Subject D, in the experiment on the weavers. It can be seen that fewest picks were missed from the tenth to the twelfth week and from the eighteenth to the twenty-second week. Her poorest record—most picks missed—was made around the fourteenth week and again at the end of the experiment. What is especially interesting is that the curves for the two treatments seem to go up and down together. Certainly the variation from

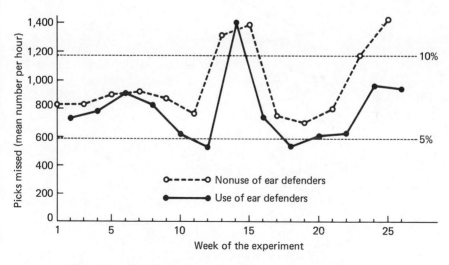

Figure 2.1 Week-to-week performance on weaving by Subject D.

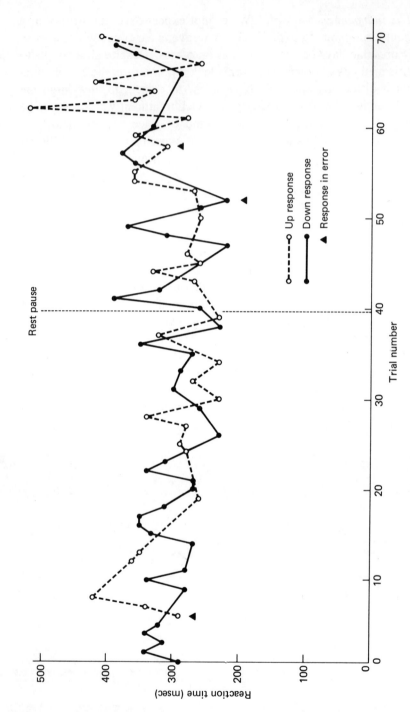

Figure 2.2 Choice reaction times on 70 consecutive trials.

Up response
Down response
Response in error

Reaction time (msec)

Trial number

Rest pause

week to week is more striking than is the difference between use or nonuse of ear defenders.

In Figure 2.2 is shown the trial-to-trial variability for a subject on a choice reaction-time task. Trials were six seconds apart; the task of the subject was to move a short lever either away from herself or toward herself, according to which of two lights came on. The two lights, of course, appeared at random. Both rapid fluctuations and longer trends may be seen over the 70 consecutive trials plotted. Shortest times occurred between about the thirtieth and fortieth trials; longest times between about the sixtieth and seventieth trials. This increase did not appear to be a matter of fatigue, as there was a rest just before the fortieth trial. In general, the very long times found were slightly above 400 msec and the very short times slightly above 200 msec, a two-to-one range.

Thus, there are second-to-second fluctuations and minute-to-minute fluctuations seen in the reaction time study. We may call them *variations in attention*; they do not appear to be related to fatigue. There are long-term shifts and trends, as shown in the record of the weaver. The ups and downs seem unrelated to temperature or humidity. However, the increase at the end might be attributable to the use of artificial (gas) light that was necessary as the experiment went into the autumn season.

The *assessment* of responses may be inconsistent even when the responses themselves are stable. Each crossing of the shuttle carrying a new pick advances a counter. But apparatus failure is not uncommon. When judgments are introduced, assessments are bound to become less consistent. Jack regarded a piece as being memorized with two successive playings without error. However, there are all kinds of little slips that border on being errors. Sometimes Jack will count them as errors, sometimes not. These inconsistencies may simply be back-and-forth fluctuations from day to day. On the other hand, they may represent trends, e.g., as his standards of accuracy change. Jack may become increasingly strict in what he calls an *error* as the experiment progresses.

Task Differences

Not only is memorizing the same piece of piano music simultaneously by two different methods an ideal that cannot be attained, but also the same piece cannot be memorized by two different methods one after the other. When it is memorized, it is memorized. There are some experiments, then, that not only require that different treatments occur at different times but also that the tasks be different. This is quite a large departure from the ideal experiment. In the case of Jack's pieces, how can he be sure that they are of the same difficulty? In any experiment using the same subject in which learning behavior is studied, there must be such task differences in treatments.

Sequence Effects

In Jack's experiment, the bad example was for him to learn the two pieces by the part method first and then the two pieces by the whole method. We already know that any variables, as just described, that change with time could affect his performance. However, there are also other possible effects related to position of a treatment within the sequence. The possible effects of one treatment on a later treatment are called *sequence effects*, *order effects*, *transfer effects*, or *carryover effects*. They may be positive or negative, general or quite specific. Jack's use of the part method on one piece might have a positive effect on his practice with the whole method on the next, because of general practice in techniques of memorizing or simply because of his getting used to the routine of the experiment. Or it might have a negative effect, because the techniques of memorizing short sections may interfere with the memorizing of large units, or simply because Jack has become tired of memorizing.

Experimenter Bias

In the early days of the automobile, there used to be a joke in the form of a riddle: *Question*—What is the most important nut in an automobile? *Answer*—The nut that holds the steering wheel. In the same spirit, we may ask a riddle: *Question*—What is the most dangerous threat to the validity of an experiment? *Answer*—The experimenter.

If the experimenter has any preference for the direction of results in an experiment, for which treatment is better, he is in a position to make the experiment come out that way. Yoko knew better than to have anything to do with working out the sequences of brands. She wanted to be sure that there was no way of guessing which brand she was tasting on any given morning. Jack was not as careful as he should have been in assigning pieces to each of his methods. He found pairs that seemed of equal difficulty to him to be practiced one for each method. He then proceeded to place them in the sequence himself. He might have unconsciously put the harder pieces of each pair in the whole method—if he wanted the part method to come out better.

Again, in making his assessments of when he had learned a piece,

Jack may not simply have been inconsistent, as was mentioned before. He again might have slightly favored one of the methods. So he would give the benefit of the doubt to the part method, while insisting on quite perfect playing for the whole method in deciding whether he had played two perfect renditions of a piece.

Since it is likely that the experimenters on the weavers expected the ear defenders to improve performance, it is also likely that they conveyed this attitude to the workers. This would result in cooperative weavers trying harder when wearing ear defenders.

One of the most insidious results of such bias is in deciding to eliminate certain data because of unusually bad conditions such as street noise. The experimenter's judgment of what conditions are unusually bad unfortunately may be swayed by expectations. Consequently, the same noise would be regarded as bad when one treatment was in effect, but not bad when they occurred during the other treatment.

Even the accuracy of recording of data has been found to depend on bias. For example, it has been shown that the recording of responses in an experiment on extrasensory perception will have errors favoring extrasensory perception if that is the recorder's belief. Apparently those who do not believe in extrasensory perception were not affected by their prejudice (Kennedy, 1939). A thorough study of the entire problem is presented in the book *Experimenter Effects in Behavioral Research* (Rosenthal, 1976).

IMPROVING INTERNAL VALIDITY BY PRIMARY CONTROLS

The term *control* is used in this book to refer to any of the ways used to improve experiments, to make them more nearly approach the perfect experiment. We are here concerned with threats to internal validity and for controls against these threats. If the controls are not adequate for an experiment, there are two ways in which its internal validity may suffer. They are called *unreliability* and *systematic confounding*. We will see how the use of *primary controls*, those which are of concern in any experiment, whatever its design, may improve reliability and reduce the chances of systematic confounding.

Unreliability

Let us suppose that the experiment on the weavers was conducted on only (what was to become) the eleventh and fourteenth week of the experiment. To keep prejudice out of the experiment, a coin was tossed to choose the week for use of ear defenders. It came up "heads," so Week 14 was chosen. Now, let us look at Figure 2.1. Performance for Subject D was poor that week: She missed more than 1,400 picks per hour. On the other hand, Week 11 was one of her good weeks; she missed fewer than 800 picks per hour. Certainly this was a fair experiment, but it was far too short. Two other weeks would give a different result, and so on. The two-week experiment is very far from the infinite experiment; the 26-week experiment was much closer, and we have good reason to believe that another 26-week experiment would give about the same results. Of course, if there are smaller ups and downs, it will not be necessary to have as many trials. The *reliability* of an average, i.e., a value of the dependent variable for one of the treatments, refers to how *stable* it would be if the experiment were done a second time, a third time, a fourth time, and so on. If it is highly reliable, about the same value would be obtained each time the experiment was repeated.

Adequate Sampling of Behavior   To obtain high reliability, then, there must be an adequate sampling of behavior relative to its variability. For a reaction-time study, as may be seen in Figure 2.2, this will require many trials, 50 or 100 or even more. Performance fluctuates greatly from trial to trial. In Jack Mozart's experiment on memorizing piano pieces, far fewer trials would be necessary, because so much behavior goes into each of the assessments, making for greater stability. An adequate sampling of behavior is thus required—relative to fluctuations in assessments.

Reduction in Variability over Time   If one way of improving reliability is to have more trials, the other way is to reduce variability. To a large extent, this is what is accomplished by *planned manipulation and documentation*, as discussed in Chapter 1. Procedures are followed exactly and on schedule, and relevant information is recorded, not left to the fallibility of memory. Beyond making sure that procedures are followed, fluctuations are further reduced as *experimental precision* is introduced.

Everything that is done to make an experiment more precise will reduce time fluctuation. Where possible, automatic methods are used, so that variations in the experimenter's behavior will not add to the sum total of time variation. A choice-reaction time experiment could be done by having the experimenter say "ready" about one second before he provides the signal for response. He will, of course, not be absolutely consistent in either his timing or the loudness of his voice. It is much better to have a timing mechanism turn on a standard light as the warning for the subject.

Again, the experimenter *could* measure reaction time with a stopwatch—starting it when he presents the signal to the subject, stopping it when he sees the subject respond. This would really be throwing a lot of trial-to-trial variation into the experiment. Almost all of this kind of variation can be eliminated by use of an electric or electronic clock that automatically starts with the signal and automatically stops with the response. Further, there are devices made to register the time in permanent form so that possible errors in recording are prevented.

It follows naturally that everything that can be done to hold other variables constant will improve reliability. Jack Mozart would certainly not practice one day in the experiment with his windows closed and another day with his windows wide open. Assuming that he is able to maintain a comfortable temperature, it would be better for him to always keep the windows closed, to reduce traffic noises and, more important, variation in traffic noises. We have already seen how careful Yoko was to keep the temperature constant of the tomato juices she drank.

Systematic Confounding

Actually, unreliability is not the worst possible sort of internal invalidity. In principle, it can always be cured by more trials. We use the term *systematic confounding* when that cure is unavailable. In the "bad" experiment on the weavers, using ear defenders for the first 13 weeks and not using them for the next 13 weeks is a poor plan that cannot be expanded to more weeks of experimentation in any practical way. Even if the whole experiment were repeated, we would be left to wonder whether one of those long 13-week periods covered a time when time variables tended to be mostly at favorable (or unfavorable) levels. In contrast, the alternate week plan actually used could easily be expanded to as many weeks as were needed for reliability. Consequently, the "bad" plan has led to a systematic confounding of the *independent variable*, use or nonuse of ear defenders, with the *time variable*, first 13 weeks or second 13 weeks.

Systematic confounding means that each level of the independent variable is accompanied in some consistent way with one of the levels of another variable. Use of ear defenders is accompanied by the first 13 weeks; nonuse of ear defenders is accompanied by the second 13 weeks. We will not be able to tell from this experiment, whether superior performance was due to a level of the independent variable (say, use of ear plugs) or to a level of another variable (first 13 weeks). You can see why this effect is called *systematic* and why unreliability is unsystematic. Reliability is improved by more trials, because any favoritism will wash out in the long run. If there is systematic confounding in an experiment, more trials will merely lead to more of the same.

If there is no good safeguard against experimenter bias, very much the same kind of systematic effect may occur. If the experimenter in the weaver's study expected ear defenders to be better, had communicated this belief to the weaver, and had made erroneous readings of the pick counter, this would all tip the balance toward ear defenders. Here the independent variable, use or nonuse of ear defenders, would be confounded with the variable experimenter bias. One of the treatments, use of ear defenders, would be accompanied by a different level of a second variable, favorable experimenter bias, from the level of the other treatment, nonuse of ear defenders, which would be accompanied by unfavorable experimenter bias. It would not be necessary for the experimenter to be unfair every time he got a chance. Even occasional occurrence of experimenter bias would have a systematic effect, being uneven for two treatments.

EXPERIMENTAL DESIGNS TO CONTROL FOR THE TIME VARIABLE IN A SINGLE SUBJECT

In the experiments described thus far, the different treatments of the experimental variable are given to the same subject. In each case, actual experiments differed from the ideal one in that the treatments were given *at different times*. Hence, the experiment must control somehow for the large collection of factors that may be lumped together under the name the *time variable*. There are really only three ways of assigning an order of treatments for this purpose that we must consider seriously. Such designs as giving one treatment entirely first and the other entirely second, as in the case of the "bad" experiment for the weavers, are simply out of the question. The three experimental designs are those which were used in the three experiments described in Chapter 1: random order of treatments (Yoko's experiment on tomato juice testing), alternating order of treatments (the weavers), and counterbalanced order of treatments (Jack Mozart's memorizing of piano pieces). We will now take up the reasons for using each of the designs and consider how well they do control for time variation.

The Random-Order Design

This design is especially well suited to experiments where trials are brief and where many of them must be used for the sake of reliability. It is the only design possible when the subject must be kept ignorant of the treatment on a given trial, as in Yoko's experiment. As the name implies, the different treatments of the independent variable are simply given in random order, for example by throwing a die and marking down whether it comes up with an odd number of dots or an even number. If two treatments are to be compared, Treatment A can be given the assignment when the number is odd, and treatment B when it is even. If the experimenter wants to make sure of getting the same number of trials of each treatment, he may simply use the method described in Yoko's experiment, with an equal number of slips of paper to be drawn for each treatment. A more systematic approach to randomization is given in Chapter 4.

In this design, there can be no systematic confounding with the time variable, since a random order is the direct opposite of a systematic one. With more trials, reliability is improved.

The Alternating-Order Design

It can be understood why the experimenters in the weavers' study used an alternating order of treatments rather than a random order. They were not sure of how long they would be weaving the same kind of cloth and wanted an equal number of weeks with and without the ear defenders. A random order might have resulted in use of the ear defenders on 10 of the first 15 weeks. Also, of course, if there are any time trends in health of operator, condition of loom, humidity, etc., one of the two treatments might be favored by appearing more often early (or late) in the series of weeks. Thus, to be on the safe side, an alternating order of treatments seemed called for rather than a random order. An alternating-order design, then, is used when each treatment is given on a fairly large number of trials or separate periods but on too few to depend on randomization for equating the time variable. Of course, Yoko could not have used a regular order of any kind, because of possible bias.

There is one little cloud in the sky in the use of an alternating order. What if there was an event that took place every other week, such as an adjustment of the loom? This would be a systematic confounding variable, favoring the treatment occurring immediately after adjustment.

Otherwise, as in the random-order design, any effect of time variation in the experiment can be compensated for by carrying on the experiment for a longer number of weeks, since the problem would be that of unreliability rather than systematic confounding with another variable.

The Counterbalanced-Order Design

A counterbalanced order of treatments is used when the experimenter, for one reason or another, does not wish to use a large number of trials or blocks of trials. This was the case for Jack Mozart. He wished to obtain information rather quickly on the best method for memorizing rather than carry his experiment through so that it included many of the pieces he would eventually memorize. Thus, his counterbalanced order, ABBA (whole-part-part-whole), consisted of only four trials. Similarly, an experimenter who wanted to compare reaction time to the occurrence of a tone with reaction time to the occurrence of a light might use four *blocks* of trials, each block consisting of 50 trials. All the trials in each block would use either a tone or a light. A counterbalanced order of treatments would be tone-light-light-tone, in which case each part of the ABBA sequence would stand not for a single trial but for a block of trials.

Reliability, as for the other designs, depends on an adequate sampling of behavior. This might be either the amount of activity that is finally assessed or the number of brief trials in a block. What may be said concerning the control over systematic confounding with time variation? The counterbalanced order makes certain that the two treatments occur, on the average, equally early and late in the experiment. It thus controls for any trial-to-trial variation that changes evenly—i.e., linearly—with time. As was mentioned earlier, in the sequence of four trials in Jack's experiment, the two treatments occurred *on the average* on order position 2.5. If it is assumed that whatever changes in time occur do so in a straight-line manner, this design controls well for time. Thus, if between each piece played, Jack's level of attentiveness increased by four points (whatever that means), the time variable would be controlled for. Let us say that attentiveness had a value of 90 when the first piece was practiced, 94 when the second piece was practiced, 98 when the third piece was practiced, and 102 when the fourth piece was practiced. This would give an average attentiveness value of 96 to Treatment A (Trials 1 and 4) and also of 96 to Treatment B (Trials 2 and 3).

All we have to do to worry about the assumption of straight-line change in the time variable is to look at Figures 2.1 and 2.2. The changes are quite irregular and unpredictable. Thus, we could imagine attentiveness increasing by 5 points, 2 points, and 1 point between trials. Over four trials, this would now give the values 90, 95, 97, and 98. The average for Treatment A (90 and 98) is here 94, while that for Treatment B (95 and 97) is 96. The counterbalancing will not have equated the time variable for the two treatments. Because of the systematic design, the result is *systematic confounding* of the independent variable with the time variable.

HOW TO HANDLE THE TASK VARIABLE

It will be remembered that if Jack had been able to do the ideal experiment, he would have learned the same piece by the two different methods. Since that is impossible, the next best thing is to have the pieces *matched* in difficulty of learning. This is the problem faced by any experiment in which it is necessary, because of learning, to use different tasks for the different methods. Let us see how the task variable is handled by each of the three designs, including that of counterbalanced order, which was the one used by Jack.

Random Order

To begin with, many more than four pieces would have been necessary, probably too many for this design to have been practical. However, let us assume it was worthwhile doing the experiment this way. There are two strategies for handling the task variable. The first would be to decide on the 30 or 40 pieces to be memorized and then to randomize the order of pieces. Each name could be put on a card and the pieces simply chosen in that order. Thus, there would be a random order of pieces as well as a random order of treatments. The other strategy would be to match pieces in difficulty. If there were 30 pieces, the two most difficult would be paired, the next two most difficult and so on, making 15 pairs in all. Within each pair a random choice, e.g., by flipping a coin, would be made as to which to assign to each method. They could then be assigned in order of difficulty, starting with the easiest pair for the first time each of the practice methods was used, and working up. Or, instead, those pieces selected for each method could be assigned in random order. However the different pieces are assigned, the only effect of variation in difficulty would be to add to the time variable. The task variable then would simply become part of the time variable. A high degree of variability in task difficulty would require more trials to attain reliability but there would be no systematic confounding.

Alternating Order

Since this order would be used if somewhat fewer pieces were to be memorized, the experimenter would be better off using the method of random matching by pairs rather than simply assigning the whole group of pieces randomly. It would be most suitable, then, first to use the two easiest pieces, the next two, etc. Again, more trials could be added if task variability is excessive, making reliability too low. There is no systematic confounding with task difficulty.

Counterbalanced Order

With a single counterbalanced series ABBA, which Jack used, matching of pieces becomes crucial. Jack did try to have the two longer pieces assigned to each method and also the two shorter pieces. He further selected pairs that were fairly equal in difficulty. He should not have made a subjective decision as to which method to use with each piece, however. He should have assigned the two pieces of each pair randomly. In that way, he would have avoided unconsciously assigning the easier piece of each pair to his preferred method of practice. However the pieces are assigned, he cannot be sure about the level of difficulty of the task variable. Consequently, the assumption is made that the matching has been accurate. There is necessarily systematic confounding with the task variable. How serious this is depends on how good the matching assumption is. Wherever possible, the experimenter should use task material that has been premeasured. For example, even nonsense syllables have been found to differ in meaning. If two lists of nonsense syllables are to be paired, they should be equal in meaning.

THE PROBLEM OF SEQUENCE EFFECTS

There is a growing realization of the problem of sequence effects in experiments (Poulton and Freeman, 1966). We shall see why they are a basic threat to the internal validity of a single-subject experiment. The term *sequence effects* means differences in response on a given trial brought about by the occurrence of previous trials. These effects may be positive (helpful) or negative (harmful). They may be general, such as getting accustomed to the routine of the experiment or becoming fatigued on later trials. They may be specific, such as learning the trick of detecting cues that tell when a shuttle is about to run out of thread. They may be temporary, lasting only to the next trial, or long term, perhaps accumulating from trial to trial. Most often they are called *carryover* or *transfer* effects.

Uniform and Nonuniform Effects

Let us consider transfer that accumulates from trial to trial. Suppose Jack Mozart became increasingly accustomed to the routine of the experiment over his four trials, whole-part-part-whole. If the effect is *uniform* between each trial and the next, neither method will be given an advantage. For example, if the positive transfer amounts to 2 "points," there would be the following amount of aid: first whole, no aid; first part, 2 points; second part, 4 points; second whole, 6 points. In total, each method, whole or part, will receive 6 points of transfer. The effect of uniform transfer thus balances out.

On the other hand, it is often true that learning will be fastest at first and will then taper off. So the transfer from the first to the second trial might amount to 3 points, with an additional 2 points from the second to third trial, and only an additional 1 point from the third to fourth trial. If such nonuniform transfer occurs, the aid on each trial would be: first whole, no aid; first part, 3 points; second part, 5 points; second whole, 6 points. Now the whole method still benefits by 6 points over all, while the part method benefits by a total of 8 points.

In the ABBA counterbalanced order, then, there is confounding of treatment effect, A or B, with another variable, *early-or-late transfer*. Treatment A is associated with late transfer, since it benefits only on the fourth trial, and Treatment B is associated with early transfer, on the second and third trials. Internal validity is threatened to the extent that the assumption of uniform transfer is unjustified. The same line of reasoning would hold if the sequence effect was negative instead of positive, e.g., accumulated fatigue. However, now Treatment A would be favored.

With the longer sequences of trials used with random and alternating orders, the problem of nonuniform effects is not nearly as serious as with the counterbalanced order. Each treatment occurs several times early and late in the experiment. However, there is an even greater danger from asymmetrical effects that does threaten internal validity for each of the three orders. Let us go on to this sequential effect now.

Symmetrical and Asymmetrical Effects

If it could be known for an experiment on a single subject that the sequence effects are symmetrical, difficulties would be much reduced. Let's see what the word *symmetrical* means in this connection to learn why this is so. What this means is that the effect on Treatment B of preceding Treatment A is the same as the effect on Treatment A of preceding Treatment B. In Jack Mozart's experiment, this would mean that the effect on learning a piece by the whole method brought about by previous learning of a piece by the part method is the same as the effect on learning a piece by the part method brought about by previous learning of a piece by the whole method.

Suppose there was interference, i.e., negative transfer, between the two methods, which amounted to 5 points. In the whole-part-part-whole sequence, the two trials that would be subject to this effect are the second and the fourth. There would be a loss of efficiency of 5 points, then, on Trial 2, part, and on Trial 4, whole. Thus symmetrical sequence effects would balance out in the counterbalanced order used by Jack. With the longer sequences used in random and alternating order, there would be just about the same number of times that A is preceded by B as there are of B being preceded by A, so again the effects would wash out.

However, if the transfer from A to B is not the same as that from B to A, the experimenter is in trouble. To make a bad case, suppose that practice with the whole method helps one in practice with the part method but that practice with the part method is harmful to practice with the whole method. Again say that the effect may be represented by a change in efficiency of 5 points. In Jack Mozart's whole-part-part-whole sequence, the second trial (part) is increased in efficiency by 5 points; the fourth trial, whole, is diminished in efficiency by 5 points. Obviously the effect will not cancel out; there is bias in favor of the part method. There need not be opposite effects in the two directions of transfer for such bias to occur, just different amounts of effect. What we have here is a systematic confounding of the independent variable, method of practice, with the sequence variable *A to B*, or *B to A*. One of the trials of Treatment A is accompanied by level *B to A* of this variable, while one of the trials of Treatment B is accompanied by level *A to B*. The real problem is that the experimenter has no way of knowing what kind of transfer is taking place; all he has are four assessments of performance, all of which are also affected by the time variable and in some cases (as in this experiment) by the task variable.

With an alternating order, the situation is not much different. Each trial of Treatment B follows a trial of Treatment A and vice versa. If there is asymmetrical transfer, there wil be systematic confounding between the independent variable and the sequence variable *on every trial* (after the first), rather than on just half of them, as in the counterbalanced order. Again, there is no way of knowing whether asymmetrical transfer is occurring.

With a random order, on about half of the trials a treatment will be preceded by the other treatment. There is some opportunity to find whether there are sequence effects and whether they are asymmetrical. For example, a separate value of the independent variable may be obtained for each of these groups of trials: A preceded by B, A not preceded by B, B preceded by A, B not preceded by A. Any difference between the first two values shows the sequence effect *B to A*; a difference between the second two values shows the sequence effect *A to B*. With this knowledge, adjustments may be made so that a value may be obtained of the dependent variable for each treatment with sequence effects subtracted out.

Still, this discussion of sequence effects does not tell the full story of how an experiment is affected by giving both treatments to the same subject. A more general kind of effect is possible. For example, the whole method of practice might be ineffective only in the context of use of the part method. It might seem very tedious under that circumstance, because of a kind of contrast effect. This would not be the case if it were the only method used. There is no method for detecting this kind of sequence effect. None of the single-subject designs, then, is free from the possibility of asymmetrical transfer. It thus may be regarded as the most serious kind of systematic confounding of the independent variable.

Moreover, it is a *general* kind of systematic confounding. If asymmetrical sequence effects exist between two treatments, they will occur in any experiment comparing those treatments. Experimenter bias will occur only for a particular experiment and may be reversed in another experiment if the new experimenter has the opposite bias. Likewise, confounding by the time variable or the task variable in short counterbalanced sequences will differ from experiment to experiment, as will nonuniform sequence effects.

SUMMARY

It is not enough for an experiment to be well documented, with planned manipulation: Some plans are poor ones. An experiment may be evaluated according to the closeness with which it approaches one of the *perfect* experiments. It is, of course, understood that perfection is unattainable. The value of the idea of a perfect experiment is that it provides a standard for judging actual experiments and in pinpointing their faults. In this chapter, the experiments described in Chapter 1 were compared, using this kind of standard, with some poorer plans for doing the same experiments.

One of the perfect experiments is the *ideal* experiment, with the different treatments given at the same time to the same subject. It emphasizes holding other variables constant. Another of the perfect experiments is the *infinite* experiment, with trials going on forever. It emphasizes obtaining a sufficiently large amount of data. The third perfect experiment is the *completely appropriate* experiment, in which all of the circumstances are the same as those to which the results will be applied. It emphasizes that other variables should not only be held constant but also that they should be held constant at an appropriate level. In all of the comparisons that were made between the experiments originally described and their poorly planned counterparts, the original ones were designed in a way closer to one of the perfect experiments.

A limited amount of fallible evidence is obtained in an experiment. The results are taken to generalize to those which would be found in a perfect experiment. The closer an actual experiment comes to a perfect experiment, which is infallible in its plans and procedures, the better the former *represents* the latter. In all of the comparisons that were made between the experiments originally described and their poorly planned counterparts, the original ones were closer to being perfect experiments. Experiments are more or less valid according to how well they represent perfect ones. Two aspects of *validity* are distinguished. The first is called *internal* validity. The question here is whether the experiment has been devised so that it will lead to the same relation between the independent and dependent variables as would be obtained in either an ideal or in an infinite experiment, that is, without the intruding effects of other variables. An experiment has *external* validity if it has been devised so that it will lead to the same results as the *completely appropriate* experiment. An experiment lacking internal validity may be called *bad*; an experiment lacking external validity may be called *wrong*. The example given of inappropriateness was an experiment in which the level of a very important other variable was wrong. It was finally pointed out that, even when an experiment is planned and performed very well, there is no absolute guarantee that the results obtained are those which would have been found in a perfect experiment. By the same token, a poorly designed experiment may come up with a correct result. However, the odds favor the better experiment.

The stress in this chapter is on internal validity, the primary requirement for all experiments. There are several threats to internal validity. First, in the single-subject experiment, there is time variation. Because an experiment cannot be ideal, an experiment cannot be done with the same subject being given the different treatments at the same time. In fact, one trial cannot be identical with another, whether the two trials are given to the same or to different subjects. In part this is due to certain other variables that chance to be at different levels. Some of these may be identified and controlled. However, much instability from trial to trial is unavoidable. Both long-term and short-term fluctuations are the rule, rather than the exception. Some of this may be attributed to inconsistent other variables that can be recognized but not really controlled. The independent variable itself may not remain constant on different presentation of the same treatment. The dependent variable may be inconsistent both because of the behavior itself and because of how it is assessed. All these factors are lumped together and called the *time variable*.

Three threats to internal validity have been described in addition to the time variable. In some experiments, particularly where learning is a factor, different tasks must be used for the different treatments, introducing a *task variable*. In any experiment using the same subject for different treatments, there are *sequence effects* of an earlier treatment on a later one. A final threat is that of *experimenter bias* in favor of one treatment over another.

All of these threats to internal validity will have one of two consequences if they are not dealt with effectively. One is that the experiment will be *unreliable*. If too few trials are obtained relative to the trial-to-trial variability, the results will be unreliable. We cannot be at all sure that the same results would be found if the experiment were repeated in the same way. The other consequence is that of *systematic confounding*, that the different treatments are systematically paired with different levels of one of the other variables that threaten internal validity.

The different single-subject experimental designs were discussed in how well they dealt with the threats to internal validity. Whatever design is used, certain primary controls are essential. These are planned manipulation and documentation, holding constant identified other variables, achieving good experimental precision, and obtaining an adequate sampling of behavior.

The random-order design is particularly suited for an experiment in which it is possible to have a large number of trials under each treatment. Its validity depends on a large enough number of trials to achieve high reliability. All of the possibilities of sequential confounding are eliminated—except that of asymmetrical transfer, which also remains for the other designs.

The alternating-order design is somewhat safer than the random-order design when not quite so many trials can be used. Reliability, again, depends on how adequately behavior was sampled. The experimenter must watch out for the possibility of some regularly occurring event that would favor one of the treatments over the other, giving systematic confounding.

The counterbalanced-order design is used when relatively few trials or blocks of trials are employed. Reliability depends on how well the trials or blocks sample behavior. The control of systematic trial-to-trial effects depends on the assumption that changes over time follow a straight-line course.

The designs differ in how to handle the task variable. When a counterbalanced order is used, tasks are matched. If the assumption that they are equal is wrong, the systematic confounding of the independent variable with the task variable, which is unavoidable in this design, brings about internal invalidity. When an alternating order or random order is used, and it is sufficiently long, various kinds of randomization of tasks are possible. Systematic task confounding does not occur. Variations in the task will simply reduce reliability.

Confounding also results from sequence effects. Nonuniformity of early and late transfer endangers internal validity when a counterbalanced order is used. The most difficult kind of systematic confounding to eliminate is that of sequence effects that are asymmetrical; i.e., the carryover effect of Treatment A on Treatment B is not the same as that of Treatment B on Treatment A. This is a problem for any experiment in which the same subject is given different treatments.

QUESTIONS

1. How may experiments be evaluated in addition to seeing whether they are well documented and have used planned manipulation?

2. Show how the ideal experiment can be used to tell us whether one way of doing an experiment is better than another.

3. Why did the infinite experiment serve as a better standard of perfection for Yoko's experiment on tasting tomato juice than did the ideal experiment?

4. How does internal validity relate to perfect experiments?

5. Can an experiment that represents the completely appropriate experiment poorly still be a good experiment?

6. Outline the threats to internal validity.

7. Why were task effects of more concern in Jack's experiment on memorizing piano pieces than in the other two experiments?

8. Distinguish between unreliability and systematic confounding.

9. How would you know whether to use an alternating-order design rather than a random-order design?

10. Compare systematic confounding in the counterbalanced-order design with that in the other two designs.

11. Distinguish between systematic confounding which differs from experiment to experiment on a given problem and that which occurs over all experiments on the problem.

STATISTICAL SUPPLEMENT

ESTIMATING POPULATION PARAMETERS FROM SAMPLE STATISTICS, THE MEAN AND STANDARD DEVIATION

Estimating the Population Mean

In the reaction time experiment described in the supplement to Chapter 1, the results of an actual experiment were taken to *represent* those which would be obtained in an experiment with perfect internal validity. Thus, the mean reaction time on 17 trials with the light signal represented the mean that would be found in an experiment with an *infinite* number of trials.

We are using the mean on a limited sample of trials to *infer* the mean on a large population (here infinite) of trials. The symbol of the mean over a population of assessments, such as RT's, is $\bar{M}_X$. This description of a population is called a *parameter*. Our actual calculated mean for a sample is called a *statistic*, M_X. Is the statistic M_X the best estimation that we may obtain from our sample of trials of the parameter $\bar{M}_X$? The answer—without giving the proof—is yes. Before you think that this is always the case, let us go on to the standard deviation, where it is not.

Computing the Standard Deviation

Usually, we want to know something besides the mean value of the assessments. We want to know how much *unsystematic variation* there is in the assessment from trial to trial. The usual way of measuring unsystematic variation is by computing the standard deviation. To do this, you find how far each assessment (i.e., X) is above or below the mean (M_X). Then you square each of these differences ($X - M_X$) and add them together. Next, you divide this sum by N, the number of trials. *Finally*, you take the square root of that average.

These computations are shown by the formula, using the symbol σ_X (small Greek letter *sigma*) for the standard deviation.

$$\sigma_X = \sqrt{\frac{\Sigma (X - M_X)^2}{N}} \qquad \text{(Formula 2.1)}$$

This can be shortened a little by using small x to stand for ($X - M_X$). Thus the formula is:

$$\sigma_X = \sqrt{\frac{\Sigma x^2}{N}} \qquad \text{(Formula 2.1A)}$$

Let us present the Treatment A data from the Chapter 1 supplement and, at the same time, go through the computations shown in the formula for σ_X.

Trial	X	M_X	$X - M_X$ or x	x^2
1	223	185	+38	+1444
2	184	185	−1	+1
3	209	185	+24	+576
4	183	185	−2	+4
5	180	185	−5	+25
6	168	185	−17	+289
7	215	185	+30	+900
8	172	185	−13	+169
9	200	185	+15	+225
10	191	185	+6	+36
11	197	185	+12	+144
12	188	185	+3	+9
13	174	185	−11	+121
14	176	185	−9	+81
15	155	185	−30	+900
16	165	185	−20	+400
17	163	185	−22	+484

$$\Sigma x^2 = +5808$$

Since

$$\sigma_A = \sqrt{\frac{\Sigma x_A^2}{N}}$$

here

$$\sigma_A = \sqrt{\frac{5808}{17}} = \sqrt{341.7}$$

$$= 18.5 \text{ msec}$$

(The square root may be found in Statistical Table 1 at the end of this supplement.)

Estimating the Population Standard Deviation

For estimating the population mean that would be obtained in an infinite experiment, the best value was, in fact, the sample mean. This is not the case for the standard deviation. In any group of actual trials, there will tend to be too few extremely high or extremely low assessments. Since the standard deviation is a measure of how spread out the assessments are, the value based on a sample is *biased*: it tends to be too low an estimate of the population parameter $\overline{\sigma}_X$.

A better estimate of population standard deviation is found by:

$$s_X = \sqrt{\frac{\Sigma x^2}{N-1}} \qquad \text{(Formula 2.2)}$$

or

$$\sigma_X \sqrt{\frac{N}{N-1}} \qquad \text{(Formula 2.2A)}$$

here,

$$s_A = \sqrt{\frac{5808}{16}} = \sqrt{363.0}$$

$$= 19.1 \text{ msec}$$

In some experiments, the hypothesis is that behavior is more *variable* under one treatment than under another. Thus, standard deviations rather than means are compared. If N is the same for the two treatments, the two *sigmas* may be compared. However, when the N's differ, the sigma for the treatment with the lower N is biased toward a lower estimate of the population standard deviation. Hence, s's should be compared.

The following table will help you keep the ideas and formulas straight.

	Mean	Standard Deviation
Population parameter	$\overline{M}_X = \dfrac{\Sigma X_{pop}}{N_{pop}}$	$\overline{\sigma}_X = \sqrt{\dfrac{\Sigma x_{pop}^2}{N_{pop}}}$
Sample statistic	$M_X = \dfrac{\Sigma X}{N}$	$\sigma_X = \sqrt{\dfrac{\Sigma x^2}{N}}$
Estimated population parameter	$M_X = \dfrac{\Sigma X}{N}$	$s_X = \sigma_X \sqrt{\dfrac{N}{N-1}}$ or $s_X = \sqrt{\dfrac{\Sigma x^2}{N-1}}$

PROBLEM: Compute σ_X and s_X for Treatment B.

Answer: $\sigma_B = 15.9$

$ s_B = 16.4$

STATISTICAL TABLE 1
Numbers 1–1,000 with their squares and square roots[a]

Number	Square	Square Root	Number	Square	Square Root
1	1	1.000	51	26 01	7.141
2	4	1.414	52	27 04	7.211
3	9	1.732	53	28 09	7.280
4	16	2.000	54	29 16	7.348
5	25	2.236	55	30 25	7.416
6	36	2.449	56	31 36	7.483
7	49	2.646	57	32 49	7.550
8	64	2.828	58	33 64	7.616
9	81	3.000	59	34 81	7.681
10	1 00	3.162	60	36 00	7.746
11	1 21	3.317	61	37 21	7.810
12	1 44	3.464	62	38 44	7.874
13	1 69	3.606	63	39 69	7.937
14	1 96	3.742	64	40 96	8.000
15	2 25	3.873	65	42 25	8.062
16	2 56	4.000	66	43 56	8.124
17	2 89	4.123	67	44 89	8.185
18	3 24	4.243	68	46 24	8.246
19	3 61	4.359	69	47 61	8.307
20	4 00	4.472	70	49 00	8.367
21	4 41	4.583	71	50 41	8.426
22	4 84	4.690	72	51 84	8.485
23	5 29	4.796	73	53 29	8.544
24	5 76	4.899	74	54 76	8.602
25	6 25	5.000	75	56 25	8.660
26	6 76	5.099	76	57 76	8.718
27	7 29	5.196	77	59 29	8.775
28	7 84	5.292	78	60 84	8.832
29	8 41	5.385	79	62 41	8.888
30	9 00	5.477	80	64 00	8.944
31	9 61	5.568	81	65 61	9.000
32	10 24	5.657	82	67 24	9.055
33	10 89	5.745	83	68 89	9.110
34	11 56	5.831	84	70 56	9.165
35	12 25	5.916	85	72 25	9.220
36	12 96	6.000	86	73 96	9.274
37	13 69	6.083	87	75 69	9.327
38	14 44	6.164	88	77 44	9.381
39	15 21	6.245	89	79 21	9.434
40	16 00	6.325	90	81 00	9.487
41	16 81	6.403	91	82 81	9.539
42	17 64	6.481	92	84 64	9.592
43	18 49	6.557	93	86 49	9.644
44	19 36	6.633	94	88 36	9.695
45	20 25	6.708	95	90 25	9.747
46	21 16	6.782	96	92 16	9.798
47	22 09	6.856	97	94 09	9.849
48	23 04	6.928	98	96 04	9.899
49	24 01	7.000	99	98 01	9.950
50	25 00	7.071	100	1 00 00	10.000

[a]Reproduced by permission from John H. Mueller, Karl F. Schuessler, and Herbert L. Costner, *Statistical Reasoning in Sociology*, 3rd ed. (Boston: Houghton Mifflin, 1977, Table V, pp. 516–528).

Number	Square	Square Root	Number	Square	Square Root
101	1 02 01	10.050	151	2 28 01	12.288
102	1 04 04	10.100	152	2 31 04	12.329
103	1 06 09	10.149	153	2 34 09	12.369
104	1 08 16	10.198	154	2 37 16	12.410
105	1 10 25	10.247	155	2 40 25	12.450
106	1 12 36	10.296	156	2 43 36	12.490
107	1 14 49	10.344	157	2 46 49	12.530
108	1 16 64	10.392	158	2 49 64	12.570
109	1 18 81	10.440	159	2 52 81	12.610
110	1 21 00	10.488	160	2 56 00	12.649
111	1 23 21	10.536	161	2 59 21	12.689
112	1 25 44	10.583	162	2 62 44	12.728
113	1 27 69	10.630	163	2 65 69	12.767
114	1 29 96	10.677	164	2 68 96	12.806
115	1 32 25	10.724	165	2 72 25	12.845
116	1 34 56	10.770	166	2 75 56	12.884
117	1 36 89	10.817	167	2 78 89	12.923
118	1 39 24	10.863	168	2 82 24	12.961
119	1 41 61	10.909	169	2 85 61	13.000
120	1 44 00	10.954	170	2 89 00	13.038
121	1 46 41	11.000	171	2 92 41	13.077
122	1 48 84	11.045	172	2 95 84	13.115
123	1 51 29	11.091	173	2 99 29	13.153
124	1 53 76	11.136	174	3 02 76	13.191
125	1 56 25	11.180	175	3 06 25	13.229
126	1 58 76	11.225	176	3 09 76	13.266
127	1 61 29	11.269	177	3 13 29	13.304
128	1 63 84	11.314	178	3 16 84	13.342
129	1 66 41	11.358	179	3 20 41	13.379
130	1 69 00	11.402	180	3 24 00	13.416
131	1 71 61	11.446	181	3 27 61	13.454
132	1 74 24	11.489	182	3 31 24	13.491
133	1 76 89	11.533	183	3 34 89	13.528
134	1 79 56	11.576	184	3 38 56	13.565
135	1 82 25	11.619	185	3 42 25	13.601
136	1 84 96	11.662	186	3 45 96	13.638
137	1 87 69	11.705	187	3 49 69	13.675
138	1 90 44	11.747	188	3 53 44	13.711
139	1 93 21	11.790	189	3 57 21	13.748
140	1 96 00	11.832	190	3 61 00	13.784
141	1 98 81	11.874	191	3 64 81	13.820
142	2 01 64	11.916	192	3 68 64	13.856
143	2 04 49	11.958	193	3 72 49	13.892
144	2 07 36	12.000	194	3 76 36	13.928
145	2 10 25	12.042	195	3 80 25	13.964
146	2 13 16	12.083	196	3 84 16	14.000
147	2 16 09	12.124	197	3 88 09	14.036
148	2 19 04	12.166	198	3 92 04	14.071
149	2 22 01	12.207	199	3 96 01	14.107
150	2 25 00	12.247	200	4 00 00	14.142

Number	Square	Square Root	Number	Square	Square Root
201	4 04 01	14.177	251	6 30 01	15.843
202	4 08 04	14.213	252	6 35 04	15.875
203	4 12 09	14.248	253	6 40 09	15.906
204	4 16 16	14.283	254	6 45 16	15.937
205	4 20 25	14.318	255	6 50 25	15.969
206	4 24 36	14.353	256	6 55 36	16.000
207	4 28 49	14.387	257	6 60 49	16.031
208	4 32 64	14.422	258	6 65 64	16.062
209	4 36 81	14.457	259	6 70 81	16.093
210	4 41 00	14.491	260	6 76 00	16.125
211	4 45 21	14.526	261	6 81 21	16.155
212	4 49 44	14.560	262	6 86 44	16.186
213	4 53 69	14.595	263	6 91 69	16.217
214	4 57 96	14.629	264	6 96 96	16.248
215	4 62 25	14.663	265	7 02 25	16.279
216	4 66 56	14.697	266	7 07 56	16.310
217	4 70 89	14.731	267	7 12 89	16.340
218	4 75 24	14.765	268	7 18 24	16.371
219	4 79 61	14.799	269	7 23 61	16.401
220	4 84 00	14.832	270	7 29 00	16.432
221	4 88 41	14.866	271	7 34 41	16.462
222	4 92 84	14.900	272	7 39 84	16.492
223	4 97 29	14.933	273	7 45 29	16.523
224	5 01 76	14.967	274	7 50 76	16.553
225	5 06 25	15.000	275	7 56 25	16.583
226	5 10 76	15.033	276	7 61 76	16.613
227	5 15 29	15.067	277	7 67 29	16.643
228	5 19 84	15.100	278	7 72 84	16.673
229	5 24 41	15.133	279	7 78 41	16.703
230	5 29 00	15.166	280	7 84 00	16.733
231	5 33 61	15.199	281	7 89 61	16.763
232	5 38 24	15.232	282	7 95 24	16.793
233	5 42 89	15.264	283	8 00 89	16.823
234	5 47 56	15.297	284	8 06 56	16.852
235	5 52 25	15.330	285	8 12 25	16.882
236	5 56 96	15.362	286	8 17 96	16.912
237	5 61 69	15.395	287	8 23 69	16.941
238	5 66 44	15.427	288	8 29 44	16.971
239	5 71 21	15.460	289	8 35 21	17.000
240	5 76 00	15.492	290	8 41 00	17.029
241	5 80 81	15.524	291	8 46 81	17.059
242	5 85 64	15.556	292	8 52 64	17.088
243	5 90 49	15.588	293	8 58 49	17.117
244	5 95 36	15.620	294	8 64 36	17.146
245	6 00 25	15.652	295	8 70 25	17.176
246	6 05 16	15.684	296	8 76 16	17.205
247	6 10 09	15.716	297	8 82 09	17.234
248	6 15 04	15.748	298	8 88 04	17.263
249	6 20 01	15.780	299	8 94 01	17.292
250	6 25 00	15.811	300	9 00 00	17.321

Number	Square	Square Root	Number	Square	Square Root
301	9 06 01	17.349	351	12 32 01	18.735
302	9 12 04	17.378	352	12 39 04	18.762
303	9 18 09	17.407	353	12 46 09	18.788
304	9 24 16	17.436	354	12 53 16	18.815
305	9 30 25	17.464	355	12 60 25	18.841
306	9 36 36	17.493	356	12 67 36	18.868
307	9 42 49	17.521	357	12 74 49	18.894
308	9 48 64	17.550	358	12 81 64	18.921
309	9 54 81	17.578	359	12 88 81	18.947
310	9 61 00	17.607	360	12 96 00	18.974
311	9 67 21	17.635	361	13 03 21	19.000
312	9 73 44	17.664	362	13 10 44	19.026
313	9 79 69	17.692	363	13 17 69	19.053
314	9 85 96	17.720	364	13 24 96	19.079
315	9 92 25	17.748	365	13 32 25	19.105
316	9 98 56	17.776	366	13 39 56	19.131
317	10 04 89	17.804	367	13 46 89	19.157
318	10 11 24	17.833	368	13 54 24	19.183
319	10 17 61	17.861	369	13 61 61	19.209
320	10 24 00	17.889	370	13 69 00	19.235
321	10 30 41	17.916	371	13 76 41	19.261
322	10 36 84	17.944	372	13 83 84	19.287
323	10 43 29	17.972	373	13 91 29	19.313
324	10 49 76	18.000	374	13 98 76	19.339
325	10 56 25	18.028	375	14 06 25	19.363
326	10 62 76	18.055	376	14 13 76	19.391
327	10 69 29	18.083	377	14 21 29	19.416
328	10 75 84	18.111	378	14 28 84	19.442
329	10 82 41	18.138	379	14 36 41	19.468
330	10 89 00	18.166	380	14 44 00	19.494
331	10 95 61	18.193	381	14 51 61	19.519
332	11 02 24	18.221	382	14 59 24	19.545
333	11 08 89	18.248	383	14 66 89	19.570
334	11 15 56	18.276	384	14 74 56	19.596
335	11 22 25	18.303	385	14 82 25	19.621
336	11 28 96	18.330	386	14 89 96	19.647
337	11 35 69	18.358	387	14 97 69	19.672
338	11 42 44	18.385	388	15 05 44	19.698
339	11 49 21	18.412	389	15 13 21	19.723
340	11 56 00	18.439	390	15 21 00	19.748
341	11 62 81	18.466	391	15 28 81	19.774
342	11 69 64	18.493	392	15 36 64	19.799
343	11 76 49	18.520	393	15 44 49	19.824
344	11 83 36	18.547	394	15 52 36	19.849
345	11 90 25	18.574	395	15 60 25	19.875
346	11 97 16	18.601	396	15 68 16	19.900
347	12 04 09	18.628	397	15 76 09	19.925
348	12 11 04	18.655	398	15 84 04	19.950
349	12 18 01	18.682	399	15 92 01	19.975
350	12 25 00	18.708	400	16 00 00	20.000

Number	Square	Square Root	Number	Square	Square Root
401	16 08 01	20.025	451	20 34 01	21.237
402	16 16 04	20.050	452	20 43 04	21.260
403	16 24 09	20.075	453	20 52 09	21.284
404	16 32 16	20.100	454	20 61 16	21.307
405	16 40 25	20.125	455	20 70 25	21.331
406	16 48 36	20.149	456	20 79 36	21.354
407	16 56 49	20.174	457	20 88 49	21.378
408	16 64 64	20.199	458	20 97 64	21.401
409	16 72 81	20.224	459	21 06 81	21.424
410	16 81 00	20.248	460	21 16 00	21.448
411	16 89 21	20.273	461	21 25 21	21.471
412	16 97 44	20.298	462	21 34 44	21.494
413	17 05 69	20.322	463	21 43 69	21.517
414	17 13 96	20.347	464	21 52 96	21.541
415	17 22 25	20.372	465	21 62 25	21.564
416	17 30 56	20.396	466	21 71 56	21.587
417	17 38 89	20.421	467	21 80 89	21.610
418	17 47 24	20.445	468	21 90 24	21.633
419	17 55 61	20.469	469	21 99 61	21.656
420	17 64 00	20.494	470	22 09 00	21.679
421	17 72 41	20.518	471	22 18 41	21.703
422	17 80 84	20.543	472	22 27 84	21.726
423	17 89 29	20.567	473	22 37 29	21.749
424	17 97 76	20.591	474	22 46 76	21.772
425	18 06 25	20.616	475	22 56 25	21.794
426	18 14 76	20.640	476	22 65 76	21.817
427	18 23 29	20.664	477	22 75 29	21.840
428	18 31 84	20.688	478	22 84 84	21.863
429	18 40 41	20.712	479	22 94 41	21.886
430	18 49 00	20.736	480	23 04 00	21.909
431	18 57 61	20.761	481	23 13 61	21.932
432	18 66 24	20.785	482	23 23 24	21.954
433	18 74 89	20.809	483	23 32 89	21.977
434	18 83 56	20.833	484	23 42 56	22.000
435	18 92 25	20.857	485	23 52 25	22.023
436	19 00 96	20.881	486	23 61 96	22.045
437	19 09 69	20.905	487	23 71 69	22.068
438	19 18 44	20.928	488	23 81 44	22.091
439	19 27 21	20.952	489	23 91 21	22.113
440	19 36 00	20.976	490	24 01 00	22.136
441	19 44 81	21.000	491	24 10 81	22.159
442	19 53 64	21.024	492	24 20 64	22.181
443	19 62 49	21.048	493	24 30 49	22.204
444	19 71 36	21.071	494	24 40 36	22.226
445	19 80 25	21.095	495	24 50 25	22.249
446	19 89 16	21.119	496	24 60 16	22.271
447	19 98 09	21.142	497	24 70 09	22.293
448	20 07 04	21.166	498	24 80 04	22.316
449	20 16 01	21.190	499	24 90 01	22.338
450	20 25 00	21.213	500	25 00 00	22.361

Number	Square	Square Root	Number	Square	Square Root
501	25 10 01	22.383	551	30 36 01	23.473
502	25 20 04	22.405	552	30 47 04	23.495
503	25 30 09	22.428	553	30 58 09	23.516
504	25 40 16	22.450	554	30 69 16	23.537
505	25 50 25	22.472	555	30 80 25	23.558
506	25 60 36	22.494	556	30 91 36	23.580
507	25 70 49	22.517	557	31 02 49	23.601
508	25 80 64	22.539	558	31 13 64	23.622
509	25 90 81	22.561	559	31 24 81	23.643
510	26 01 00	22.583	560	31 36 00	23.664
511	26 11 21	22.605	561	31 47 21	23.685
512	26 21 44	22.627	562	31 58 44	23.707
513	26 31 69	22.650	563	31 69 69	23.728
514	26 41 96	22.672	564	31 80 96	23.749
515	26 52 25	22.694	565	31 92 25	23.770
516	26 62 56	22.716	566	32 03 56	23.791
517	26 72 89	22.738	567	32 14 89	23.812
518	26 83 24	22.760	568	32 26 24	23.833
519	26 93 61	22.782	569	32 37 61	23.854
520	27 04 00	22.804	570	32 49 00	23.875
521	27 14 41	22.825	571	32 60 41	23.896
522	27 24 84	22.847	572	32 71 84	23.917
523	27 35 29	22.869	573	32 83 29	23.937
524	27 45 76	22.891	574	32 94 76	23.958
525	27 56 25	22.913	575	33 06 25	23.979
526	27 66 76	22.935	576	33 17 76	24.000
527	27 77 29	22.956	577	33 29 29	24.021
528	27 87 84	22.978	578	33 40 84	24.042
529	27 98 41	23.000	579	33 52 41	24.062
530	28 09 00	23.022	580	33 64 00	24.083
531	28 19 61	23.043	581	33 75 61	24.104
532	28 30 24	23.065	582	33 87 24	24.125
533	28 40 89	23.087	583	33 98 89	24.145
534	28 51 56	23.108	584	34 10 56	24.166
535	28 62 25	23.130	585	34 22 25	24.187
536	28 72 96	23.152	586	34 33 96	24.207
537	28 83 69	23.173	587	34 45 69	24.228
538	28 94 44	23.195	588	34 57 44	24.249
539	29 05 21	23.216	589	34 69 21	24.269
540	29 16 00	23.238	590	34 81 00	24.290
541	29 26 81	23.259	591	34 92 81	24.310
542	29 37 64	23.281	592	35 04 64	24.331
543	29 48 49	23.302	593	35 16 49	24.352
544	29 59 36	23.324	594	35 28 36	24.372
545	29 70 25	23.345	595	35 40 25	24.393
546	29 81 16	23.367	596	35 52 16	24.413
547	29 92 09	23.388	597	35 64 09	24.434
548	30 03 04	23.409	598	35 76 04	24.454
549	30 14 01	23.431	599	35 88 01	24.474
550	30 25 00	23.452	600	36 00 00	24.495

Number	Square	Square Root	Number	Square	Square Root
601	36 12 01	24.515	651	42 38 01	25.515
602	36 24 04	24.536	652	42 51 04	25.534
603	36 36 09	24.556	653	42 64 09	25.554
604	36 48 16	24.576	654	42 77 16	25.573
605	36 60 25	24.597	655	42 90 25	25.593
606	36 72 36	24.617	656	43 03 36	25.612
607	36 84 49	24.637	657	43 16 49	25.632
608	36 96 64	24.658	658	43 29 64	25.652
609	37 08 81	24.678	659	43 42 81	25.671
610	37 21 00	24.698	660	43 56 00	25.690
611	37 33 21	24.718	661	43 69 21	25.710
612	37 45 44	24.739	662	43 82 44	25.729
613	37 57 69	24.759	663	43 95 69	25.749
614	37 69 96	24.779	664	44 08 96	25.768
615	37 82 25	24.799	665	44 22 25	25.788
616	37 94 56	24.819	666	44 35 56	25.807
617	38 06 89	24.839	667	44 48 89	25.826
618	38 19 24	24.860	668	44 62 24	25.846
619	38 31 61	24.880	669	44 75 61	25.865
620	38 44 00	24.900	670	44 89 00	25.884
621	38 56 41	24.920	671	45 02 41	25.904
622	38 68 84	24.940	672	45 15 84	25.923
623	38 81 29	24.960	673	45 29 29	25.942
624	38 93 76	24.980	674	45 42 76	25.962
625	39 06 25	25.000	675	45 56 25	25.981
626	39 18 76	25.020	676	45 69 76	26.000
627	39 31 29	25.040	677	45 83 29	26.019
628	39 43 84	25.060	678	45 96 84	26.038
629	39 56 41	25.080	679	46 10 41	26.058
630	39 69 00	25.100	680	46 24 00	26.077
631	39 81 61	25.120	681	46 37 61	26.096
632	39 94 24	25.140	682	46 51 24	26.115
633	40 06 89	25.159	683	46 64 89	26.134
634	40 19 56	25.179	684	46 78 56	26.153
635	40 32 25	25.199	685	46 92 25	26.173
636	40 44 96	25 219	686	47 05 96	26.192
637	40 57 69	25.239	687	47 19 69	26.211
638	40 70 44	25.259	688	47 33 44	26.230
639	40 83 21	25.278	689	47 47 21	26.249
640	40 96 00	25.298	690	47 61 00	26.268
641	41 08 81	25.318	691	47 74 81	26.287
642	41 21 64	25.338	692	47 88 64	26.306
643	41 34 49	25.357	693	48 02 49	26.325
644	41 47 36	25.377	694	48 16 36	26.344
645	41 60 25	25.397	695	48 30 25	26.363
646	41 73 16	25.417	696	48 44 16	26.382
647	41 86 09	25.436	697	48 58 09	26.401
648	41 99 04	25.456	698	48 72 04	26.420
649	42 12 01	25.475	699	48 86 01	26.439
650	42 25 00	25.495	700	49 00 00	26.458

Number	Square	Square Root	Number	Square	Square Root
701	49 14 01	26.476	751	56 40 01	27.404
702	49 28 04	26.495	752	56 55 04	27.423
703	49 42 09	26.514	753	56 70 09	27.441
704	49 56 16	26.533	754	56 85 16	27.459
705	49 70 25	26.552	755	57 00 25	27.477
706	49 84 36	26.571	756	57 15 36	27.495
707	49 98 49	26.589	757	57 30 49	27.514
708	50 12 64	26.608	758	57 45 64	27.532
709	50 26 81	26.627	759	57 60 81	27.550
710	50 41 00	26.646	760	57 76 00	27.568
711	50 55 21	26.665	761	57 91 21	27.586
712	50 69 44	26.683	762	58 06 44	27.604
713	50 83 69	26.702	763	58 21 69	27.622
714	50 97 96	26.721	764	58 36 96	27.641
715	51 12 25	26.739	765	58 52 25	27.659
716	51 26 56	26.758	766	58 67 56	27.677
717	51 40 89	26.777	767	58 82 89	27.695
718	51 55 24	26.796	768	58 98 24	27.713
719	51 69 61	26.814	769	59 13 61	27.731
720	51 84 00	26.833	770	59 29 00	27.749
721	51 98 41	26.851	771	59 44 41	27.767
722	52 12 84	26.870	772	59 59 84	27.785
723	52 27 29	26.889	773	59 75 29	27.803
724	52 41 76	26.907	774	59 90 76	27.821
725	52 56 25	26.926	775	60 06 25	27.839
726	52 70 76	26.944	776	60 21 76	27.857
727	52 85 29	26.963	777	60 37 29	27.875
728	52 99 84	26.981	778	60 52 84	27.893
729	53 14 41	27.000	779	60 68 41	27.911
730	53 29 00	27.019	780	60 84 00	27.928
731	53 43 61	27.037	781	60 99 61	27.946
732	53 58 24	27.055	782	61 15 24	27.964
733	53 72 89	27.074	783	61 30 89	27.982
734	53 87 56	27.092	784	61 46 56	28.000
735	54 02 25	27.111	785	61 62 25	28.018
736	54 16 96	27.129	786	61 77 96	28.036
737	54 31 69	27.148	787	61 93 69	28.054
738	54 46 44	27.166	788	62 09 44	28.071
739	54 61 21	27.185	789	62 25 21	28.089
740	54 76 00	27.203	790	62 41 00	28.107
741	54 90 81	27.221	791	62 56 81	28.125
742	55 05 64	27.240	792	62 72 64	28.142
743	55 20 49	27.258	793	62 88 49	28.160
744	55 35 36	27.276	794	63 04 36	28.178
745	55 50 25	27.295	795	63 20 25	28.196
746	55 65 16	27.313	796	63 36 16	28.213
747	55 80 09	27.331	797	63 52 09	28.231
748	55 95 04	27.350	798	63 68 04	28.249
749	56 10 01	27.368	799	63 84 01	28.267
750	56 25 00	27.386	800	64 00 00	28.284

Number	Square	Square Root	Number	Square	Square Root
801	64 16 01	28.302	851	72 42 01	29.172
802	64 32 04	28.320	852	72 59 04	29.189
803	64 48 09	28.337	853	72 76 09	29.206
804	64 64 16	28.355	854	72 93 16	29.223
805	64 80 25	28.373	855	73 10 25	29.240
806	64 96 36	28.390	856	73 27 36	29.257
807	65 12 49	28.408	857	73 44 49	29.275
808	65 28 64	28.425	858	73 61 64	29.292
809	65 44 81	28.443	859	73 78 81	29.309
810	65 61 00	28.460	860	73 96 00	29.326
811	65 77 21	28.478	861	74 13 21	29.343
812	65 93 44	28.496	862	74 30 44	29.360
813	66 09 69	28.513	863	74 47 69	29.377
814	66 25 96	28.531	864	74 64 96	29.394
815	66 42 25	28.548	865	74 82 25	29.411
816	66 58 56	28.566	866	74 99 56	29.428
817	66 74 89	28.583	867	75 16 89	29.445
818	66 91 24	28.601	868	75 34 24	29.462
819	67 07 61	28.618	869	75 51 61	29.479
820	67 24 00	28.636	870	75 69 00	29.496
821	67 40 41	28.653	871	75 86 41	29.513
822	67 56 84	28.671	872	76 03 84	29.530
823	67 73 29	28.688	873	76 21 29	29.547
824	67 89 76	28.705	874	76 38 76	29.563
825	68 06 25	28.723	875	76 56 25	29.580
826	68 22 76	28.740	876	76 73 76	29.597
827	68 39 29	28.758	877	76 91 29	29.614
828	68 55 84	28.775	878	77 08 84	29.631
829	68 72 41	28.792	879	77 26 41	29.648
830	68 89 00	28.810	880	77 44 00	29.665
831	69 05 61	28.827	881	77 61 61	29.682
832	69 22 24	28.844	882	77 79 24	29.698
833	69 38 89	28.862	883	77 96 89	29.715
834	69 55 56	28.879	884	78 14 56	29.732
835	69 72 25	28.896	885	78 32 25	29.749
836	69 88 96	28.914	886	78 49 96	29.766
837	70 05 69	28.931	887	78 67 69	29.783
838	70 22 44	28.948	888	78 85 44	29.799
839	70 39 21	28.965	889	79 03 21	29.816
840	70 56 00	28.983	890	79 21 00	29.833
841	70 72 81	29.000	891	79 38 81	29.850
842	70 89 64	29.017	892	79 56 64	29.866
843	71 06 49	29.034	893	79 74 49	29.883
844	71 23 36	29.052	894	79 92 36	29.900
845	71 40 25	29.069	895	80 10 25	29.916
846	71 57 16	29.086	896	80 28 16	29.933
847	71 74 09	29.103	897	80 46 09	29.950
848	71 91 04	29.120	898	80 64 04	29.967
849	72 08 01	29.138	899	80 82 01	29.983
850	72 25 00	29.155	900	81 00 00	30.000

Number	*Square*	*Square Root*	*Number*	*Square*	*Square Root*
901	81 18 01	30.017	951	90 44 01	30.838
902	81 36 04	30.033	952	90 63 04	30.854
903	81 54 09	30.050	953	90 82 09	30.871
904	81 72 16	30.067	954	91 01 16	30.887
905	81 90 25	30.083	955	91 20 25	30.903
906	82 08 36	30.100	956	91 39 36	30.919
907	82 26 49	30.116	957	91 58 49	30.935
908	82 44 64	30.133	958	91 77 64	30.952
909	82 62 81	30.150	959	91 96 81	30.968
910	82 81 00	30.166	960	92 16 00	30.984
911	82 99 21	30.183	961	92 35 21	31.000
912	83 17 44	30.199	962	92 54 44	31.016
913	83 35 69	30.216	963	92 73 69	31.032
914	83 53 96	30.232	964	92 92 96	31.048
915	83 72 25	30.249	965	93 12 25	31.064
916	83 90 56	30.265	966	93 31 56	31.081
917	84 08 89	30.282	967	93 50 89	31.097
918	84 27 24	30.299	968	93 70 24	31.113
919	84 45 61	30.315	969	93 89 61	31.129
920	84 64 00	30.332	970	94 09 00	31.145
921	84 82 41	30.348	971	94 28 41	31.161
922	85 00 84	30.364	972	94 47 84	31.177
923	85 19 29	30.381	973	94 67 29	31.193
924	85 37 76	30.397	974	94 86 76	31.209
925	85 56 25	30.414	975	95 06 25	31.225
926	85 74 76	30.430	976	95 25 76	31.241
927	85 93 29	30.447	977	95 45 29	31.257
928	86 11 84	30.463	978	95 64 84	31.273
929	86 30 41	30.480	979	95 84 41	31.289
930	86 49 00	30.496	980	96 04 00	31.305
931	86 67 61	30.512	981	96 23 61	31.321
932	86 86 24	30.529	982	96 43 24	31.337
933	87 04 89	30.545	983	96 62 89	31.353
934	87 23 56	30.561	984	96 82 56	31.369
935	87 42 25	30.578	985	97 02 25	31.385
936	87 60 96	30.594	986	97 21 96	31.401
937	87 79 69	30.610	987	97 41 69	31.417
938	87 98 44	30.627	988	97 61 44	31.432
939	88 17 21	30.643	989	97 81 21	31.448
940	88 36 00	30.659	990	98 01 00	31.464
941	88 54 81	30.676	991	98 20 81	31.480
942	88 73 64	30.692	992	98 40 64	31.496
943	88 92 49	30.708	993	98 60 49	31.512
944	89 11 36	30.725	994	98 80 36	31.528
945	89 30 25	30.741	995	99 00 25	31.544
946	89 49 16	30.757	996	99 20 16	31.559
947	89 68 09	30.773	997	99 40 09	31.575
948	89 87 04	30.790	998	99 60 04	31.591
949	90 06 01	30.806	999	99 80 01	31.607
950	90 25 00	30.822	1000	100 00 00	31.623

EXPERIMENTS THAT "IMPROVE" THE REAL WORLD

Why should a highly trained pilot crash his commercial jet airliner during a routine approach to an airport on a clear, still night? That is the question asked by Conrad Kraft and Charles Elworth (1969), two experimental psychologists with the Boeing Company. But they were interested in more than a single crash. Astonishingly, almost one out of five major aircraft accidents occurs "during seemingly safe night visual approaches" (p. 2).

In order to answer this question—so that a start could be made toward solving the practical problem—Kraft and Elworth examined the statistics on accidents during night visual approaches more closely to find a clue. They believed that they found one: Airports that were of somewhat lower elevation than the city that stretched out beyond got more than their share of serious accidents. An example would be an airport that lies at the edge of a large lake. The approach is over the water to an airport only slightly higher than the water level. The city beyond slopes up gradually, away from the airport.

This led to their experimental hypothesis: Pilots use a method of visual approach that gives a good path if the city beyond is flat but too low a path if the city has an upward slope. A satisfactory experiment to test this hypothesis cannot be done by having a pilot make night landings at two different airports, one with a flat city beyond and one with a sloped city beyond. The main reason that such an experiment, one that *duplicates* the real world, would be unsatisfactory is that there would be too many variables other than the slope that would differ. The pattern of lights would not be the same, the clearness of the air and the wind flow could differ, etc. Further, there is the problem of danger. Even while making an approach using the visual scene below, pilots sometimes consult their instruments just to be on the safe side. But this would weaken the experiment. What was wanted was the knowledge of how the pilot flies an entirely visual approach—using the lights below—when the light pattern was on a flat surface and on a sloped surface.

An experiment that duplicates the real world would not do. What was needed was an experiment that *"improves"* it (for the purposes of the experiment, that is). In this case, it would be one in which the experimenters could control the pattern of lights, the visibility, and the wind flow, and could also prevent use of the altimeter without endangering the pilot. Such an experiment was carried out by Kraft and Elworth. It is the first of three experiments described in this chapter—all of which "improve" the real world of application.

These *artificial* experiments, as they may be called, are performed when those which duplicate the real world would lack internal validity. However, they raise a new question: How can we be sure that the results do apply to the real world? What are the experimental safeguards that allowed Kraft and Elworth to conclude that what they observed in their laboratory is *representative* of what occurs during a real approach to an airport? The problem of *external* validity comes to the fore. It was of little concern in the first experiments discussed in this book, since they duplicated the real world. Now, when it is necessary to improve the real world in order to obtain sufficient internal validity, external validity is thrown into doubt. Something gained, but something lost.

Most experiments that improve the real world are probably impractical for you to do at the present time. However, it would be good training for you to learn to identify problems that cannot be handled by experiments in the real world and to work out plans for artificial experiments that would be externally valid.

The questions you should be able to answer after reading this chapter fall into the following classes:

1. What are the circumstances that call for an experiment that improves the real world rather than duplicates it?
2. What are the ways in which internal validity is strengthened by this choice?
3. What are the problems of representativeness in respect to external validity?
4. How realistic should the experiments be?

AN EXPERIMENT ON NIGHT VISUAL APPROACHES

The Experimental Hypothesis

To see how Kraft and Elworth arrived at their experimental hypothesis requires some knowledge of how a pilot approaches a landing at night using visual guidance from the lights below. What must be accomplished is to descend rather steeply at first and then to make the approach more and more shallow until the plane is flying parallel to the ground just above the landing strip. This approach to a landing can be achieved if the plane descends along the curve of a large circle as if it were at the end of a

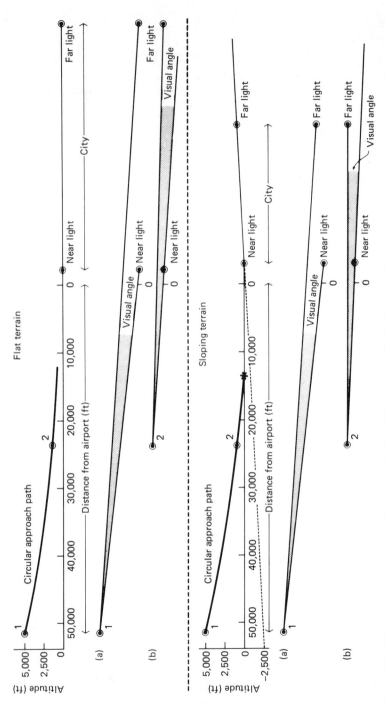

Figure 3.1 The circular approach path that results from "flying the null." *Flat Terrain.* The path with near and far reference lights on flat terrain. (a) and (b). The same visual angle is maintained in the path shown for flat terrain. *Sloping Terrain.* The path with near and far reference lights on sloping terrain. (a) and (b). The same visual angle is maintained in the path shown for sloping terrain. The * marks the place where the plane would crash if the circular approach path were continued.

pendulum that is suspended far up into the sky. You see such a circular approach path in Figure 3.1. The portion shown starts about 10 miles from the airport (a little over 50,000 feet) with the plane about 5,000 feet high.

Here is how the pilot is able to fly this path, using the visual information available. The pilot keeps looking at a pair of lights, located in the city beyond the airport, that are along the line of flight. One light is (relatively) near, and the other is far. The far light will be higher in the pilot's field of view than the near light. (You can check on that statement right now. Look at an object on the floor near where you are sitting. If you want to see an object on the floor that is farther away, you must raise your head.) If the pilot keeps the vertical distance between the near and far lights just about the same in his field of view at all points along the approach path, the plane will follow this circular path. This is called "flying the null."

Another way of putting it is to say that the pilot keeps the *visual angle* between the two lights constant. The visual angle is found by drawing a straight line from the near light to the pilot's eye and a straight line from the far light to his eye. The angle—at his eye—between these two lines is the visual angle. If the vertical distance in the pilot's field of view remains constant during the path, the visual angle has remained constant.

In the upper half of Figure 3.1, representing an approach over flat terrain, it is shown that the visual angle does remain constant as long as the pilot keeps on the desired circular path. The tiny angle at early Position 1 along the path shown in (a) is the same as the tiny angle shown in (b) at Position 2, but farther along the path. At these two points, and at all others, the vertical distance between the near and far lights in the pilot's field of view has remained constant.

However, this will give a good approach only if the lights are on a horizontal surface. In the lower half of Figure 3.1 is shown the approach path when a pilot "flys the null" using a near and a far light from a city that slopes up beyond the airport. At the first point along the circular path, the visual angle shown in (a), between the near and far lights, is the same as that shown for the flat terrain. Likewise, the same visual angle is retained farther along the path at Point 2 as shown in (b). However, that point is already near ground level. If the pilot continues to "fly the null," the plane will crash before the airport is reached. The pilot will have been the victim of an optical illusion. He has perceived the two lights to lie on a horizontal surface. The dotted line, to the left of the near light represents his placement of the horizontal surface over which he is approaching the landing. As can be seen, it is considerably lower than the real surface of the earth. So far, this is a good hypothesis that works well in drawings on paper. Now we will turn to the way in which Kraft and Elworth tested their hypothesis *in action*.

The Experiment

The Simulator   To do their experiment, Kraft and Elworth used a device called a *simulator*. It consists of a cabin in which there are the seats for the pilot and copilot, the hand and foot controls by which a plane is usually flown, a panel with all of the dials and indicators, and a view that lies ahead through the windshield. When the pilot "flies" the plane, the changes in the view presented are exactly those which would occur in real flight. Also, the changes on the dials are those which would occur. The pilot may be made to fly using only the view of the ground, only the dials, or any combination of dials and view. In this experiment, the pilot was given the view of the ground but not of the dial that shows how high he is flying, the altimeter. Figure 3.2(a) shows a test crew flying the simulator, looking ahead to the view of a city. Figure 3.2(b) shows the airport and a city view from above. Actually, the presentation was even more realistic, with many colored and flashing lights. As you have prob-ably guessed, such a simulator is a very complex device that is pro-grammed by means of a computer. In addition to the functions already described, the simulator records the ways in which the pilot moves the controls and the changes in the readings that result on the dials. For this study, a continuous record was made of the readings on the altimeter, which were withheld from the pilot.

Method   *Task*: The pilot flew descent paths to an "airport" as shown in Figure 3.2, the "city" being presented either as flat or as on terrain that slopes up from the airport at a 3° angle. [This is the same slope that is shown in the lower half of Figure 3.1.] He was allowed to fly the descent path as he wished, but to try to fly 180 mph at an altitude of 5,000 feet at 10 miles from the airport, and at 120 mph at an altitude of 1,250 feet at 4.5 miles from the airport. [These are the points along the flight path numbered 1 and 2 in Figure 3.1.] You should again note that the pilot was required to use only the visual information through the windshield; he was not allowed to use the altimeter to find out how high he was above the ground.

In addition to the continuous registration of the "actual" altitude being flown, a tape recording was made of the pilot's voice, in which he

(a)

(b)

Figure 3.2 The simulator is shown in (a); a view of the lighted city is shown in (b). Reproduced by permission of Dr. K. L. Craft and Dr. C. L. Elworth, and the Boeing Co., from "Night Visual Approaches," *Boeing Airliner*, 1969 (March-April), 2–4.

estimated the altitude periodically. In order to make the pilot's task more demanding, as in real flight, he was required to locate and report on other "aircraft" in the field of view.

Procedure: There were ten trials, on each of which the pilot made a descent. Some other variables, in addition to the main independent variable of slope, were systematically changed from trial to trial: e.g., light patterns, darkened areas, starting altitude, and distance.

Results   Shown in Figure 3.3 are the average flight paths flown by 12 experienced pilots. The altitude scale has been expanded so differences show up more clearly. It can be seen that the approach path with the city on flat terrain is much like that predicted in Figure 3.1 (upper half), if the pilot flew the null using two reference lights on a flat terrain. The approach path with the city at a 3° angle is like that predicted in Figure 3.1 (lower half), but not so drastic; the plane does not actually "crash." Also shown in this figure are the pilots' estimates of altitude. It is seen that even with the flat terrain there was a tendency to overestimate altitude. The pilot believed he was higher above the surface than he really was. With the sloping terrain, there is a gross exaggeration. Note that at 10 miles out the pilots believed that their altitude was almost 4,000 feet when, in fact, it was less than 2,000 feet!

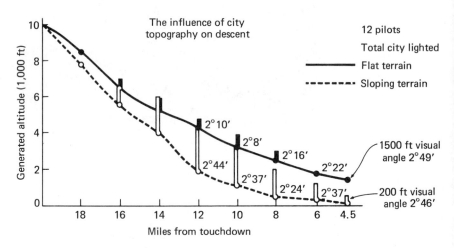

Figure 3.3 Average paths flown by pilots in the simulator. Tops of bars show average estimates. Reproduced by permission of Dr. K. L. Craft and Dr. C. L. Elworth, and the Boeing Co., from "Night Visual Approaches," *Boeing Airliner*, 1969 (March-April), 2–4.

   Discussion and conclusions   The predictions came true; with a sloping city, the pilot flew an approach path almost as if the city beyond were flat. Like other visual illusions, the effect cannot be eliminated by verbal information. "In each case, the pilots were advised immediately prior to the approach as to whether the city was sloped or flat. In other words, the pilots continued to view the city/airport light patterns, spread out over varying topography, as representatives of a flat city" (p. 4). Since there is bound to be this illusion or misperception, pilots should not rely on "flying the null," but must make continuous checks with their altimeter.

IMPROVEMENT 1: ELIMINATION OF SYSTEMATIC CONFOUNDING

Suppose that Kraft and Elworth, instead of using the simulator, had done their experiment in a way that duplicates the real world. They would first of all have selected two different airports, that differed in slope of the city beyond, to compare night visual approaches. They would, of course, have tried as well as they could to keep other possible variables at similar levels. Thus they would have tried to select similar surfaces, water or land, over which the approach is made. They would have looked for cities of about the same size, with similar patterns of lights. They would have tried to match the regions in visibility. It would have finally turned out to be impossible to find two airports that were closely matched in respect to all identifiable other variables. They would have had to choose those thought to be most important. Even if they had done very well in their matching, new differences would have appeared at the time of the trial. The amount of air traffic could not have been counted on to be the same. Instructions from the control tower would have differed, etc.

   The experiment in the real world is thus very far from the *ideal experiment*. Even where other variables are matched for the different airports, they are only at somewhat the same level, not at the same identical level as with the simulator, which "improves" the real world. As a consequence, in an experiment on two different airports there could not be very good assurance that the difference in the approach path flown to a level and to a sloped city was due to the variable of slope. The other variables, on which the two airports differ at least slightly, could have been responsible for the entire difference, could have reduced the real difference, etc. In other words, the experiment that duplicates the real world has poor internal validity because there is *systematic confounding* of other variables with the independent variable, namely slope of the city.

We have now seen one reason for conducting an artificial experiment, one that "improves" the real world. It is called for when an experiment that duplicates the real world would be *internally* invalid because of the systematic confounding by other variables. The experiment described was able to avoid this confounding by holding other variables constant.

We will now consider two other experiments that emphasize still other advantages, in respect to internal validity, of improving the real world. Unlike the experiment just described, they are make-believe examples. The planning is described in more detail than in the preceding one, to emphasize the decisions the experimenter must make. They are considerably less technical than the night landing approach experiment and so are much more the kind of thing you could do right now.

AN EXPERIMENT ON SEA RESCUE SEARCH

When there is a report of a vessel in distress or when a boat fails to return as expected, it is the task of the Coast Guard to attempt a rescue. Depending on circumstances, any or all of the following may be used: small ships (patrol boats), amphibious planes, or helicopters. In many cases, as in looking for survivors of an overturned sailboat, sophisticated devices such as radar or sonar are of no use. People must *look* over the water to try to catch a glimpse of anything floating. Let us imagine the following scene taking place between two crew members of the Coast Guard station at Santo Tomás, while holystoning the foredeck of patrol boat CG-99999.

Hawkeye: Are you looking forward to the routine training sessions coming up on rescue operations, dear?

Dionne: Routine says it all. We keep practicing the same old thing. I'm not so sure that we're even searching the right way.

Hawkeye: Whadya mean?

Dionne: Like take those clugey 7 × 50 binoculars. I'm for throwing them overboard and peering out with my naked eyes.

Hawkeye: You're putting me on.

Dionne: Let's face it. First of all you can see only an itsy-bitsy angle of the ocean with those monstrosities. It would take ages to really cover even a square mile of ocean. Second, with the ship rolling and pitching and yawing, you can't see the same spot for more than a jif. Third, you can't for the life of you tell how far away anything is; the ocean looks like it goes up instead of out. Fourth, after you have looked through them for a while your eyes revolt. Finally, they weigh a ton.

Hawkeye: Ha! That's what comes from letting the weaker sex into this man's Navy. Anyway, how about all of those old movies showing the horizon through the two big circles side-by-side?

Dionne: The Coast Guard is under the Department of Transportation, and I'll skip any comment on the second matter. Anyway, why don't we talk Commander Laughton into letting us make a test of search with and without binoculars, instead of just dragging through the training exercises?

Hawkeye: Remember when Beldell suggested that we train dolphins to do the searching, the awful way the Commander sneered at him?

Dionne: Don't worry, I'll tell it to him so roundabout he'll think he thought it up all by himself.

Hawkeye: I believe you.

Dionne: Here is what we need to do. First, you and I should each search with and without binoculars. That'll be easy; we'll only use one pair. Second, we need to put lots of targets into the water so we can be timed on enough so that one or two lucky breaks won't matter. Third, we should have targets of different sizes, like the sizes of rafts and people. Fourth, we should search at different times of day in different kinds of weather, and with targets at different distances. Fifth . . .

Hawkeye: Don't you ever get tired of listing things? Anyway, what kind of targets?

Dionne: I've got it! Inflatable rafts from a surplus store and seahorse surf blowups for people.

Hawkeye (Getting into the spirit): Yeah, pink ones for people!

Dionne: Some brown ones too.

Hawkeye (swallows): How are we going to tell whether we have already sighted one if there are to be lots of targets?

Dionne: Hmm. We'll have them dropped one at a time from a helicopter. We'll be blindfolded when this is done. After we have both sighted it, or maybe after fifteen minutes, the guys in the chopper can either catch it or harpoon it. We'll find out which works best.

Hawkeye: What'll the ship be doing?

Dionne: Cruising in tight circles, or even squares, if that's what the skipper wants to do. The only thing that's really important is that he doesn't steam away too far. Also, somebody on board who saw the drop should be plotting its bearings every half minute or so. What we should each do, when sighting the target, is to write down the bearings and the time.

Hawkeye: You've been talking so much and so fast, how are we ever going to remember all this stuff?

Dionne: Remember? None of that stuff. What if some nosey inspector wants to know how come we're not following doctrine and wants us to prove we made a fair test? We'll write everything down in some extra logs they have in ships stores. In one of them I'll write a table of contents, then a plan for the experiment, then protocol pages—we'll need a second book so we can both make entries, and finally a summary.

Hawkeye: Gosh!

We now leave this unlikely conversation and examine the

Plan of the Experiment

Since the general ideas and ways of going about the experiment have already been revealed, there is shown in Table 3.1 only the formal design of the experiment.

It was originally planned to have the target dropped in one of four sectors around the cutter, to use four different types of target two times of day, and, of course, the subjects using two different methods. By multiplication, it is seen that to have each unique combination occur just once, there would have to be 192 target drops. This plan was considered impractical by Commander Laughton and was therefore pared down. It is typical for the initial planning of an experiment to be grandiose. With the cutter moving in circles, position of drop was considered to be irrelevant; it is only important that the drop be outside the circle, on the starboard side. Further, the budding experimenters finally conceded that they could live with only two types of targets and two kinds of weather.

Even so, 48 target drops are required. The experiment is to be conducted in eight sessions, corresponding to midday (high sun) and late afternoon (low sun) conditions. For the first four sessions (A–D) Hawkeye will use the binoculars and they will switch methods for the last four sessions (E–H). Within each session, the order of the six target size-range combinations was drawn at random.

TABLE 3.1

**Design of the experiment: Search in sea rescue
operations (four of the eight sessions shown here)**

Treatments
Experiment comparing search with and without 7 × 50 binoculars. The tryouts will be made on four days in the month of October, 1977, if possible on two sunny days and two stormy days.

1. Conditions for viewing (2). 7 × 50 binoculars and unaided vision. Subject 1 (binoculars, first half). Subject 2 (binoculars, second half).
2. Weather (2). Sunny and Stormy.
3. Time of Day (2). Early (1100–1300) and Late (1600–1800).
4. Targets (2). Large (raft) and Small (surf blowup).
5. Ranges (3). 0.5 mile, 1 mile, 4 miles.

Schedule

Trial Session A (Early, Sunny)

1 Small target,	1	mile
2 Large target,	4	miles
3 Large target,	4	miles
4 Large target,	1	mile
5 Small target,	0.5	mile
6 Small target,	0.5	mile

Trial Session C (Early, Stormy)

1 Small target,	4	miles
2 Large target,	0.5	mile
3 Large target,	1	mile
4 Small target,	1	mile
5 Small target,	4	miles
6 Large target,	0.5	mile

Trial Session B (Late, Sunny)

1 Large target,	1	mile
2 Small target,	1	mile
3 Small target,	0.5	mile
4 Large target,	0.5	mile
5 Large target,	4	miles
6 Small target,	4	miles

Trial Session D (Late, Stormy)

1 Large target,	1	mile
2 Large target,	4	miles
3 Large target,	1	mile
4 Small target,	0.5	mile
5 Small target,	4	miles
6 Small target,	0.5	mile

Analyzing the Data

A systematic representation of the results is written into the laboratory notebook. The portion concerning Hawkeye is shown in Table 3.2. The very top line compares his performance with binoculars and unaided vision for the large targets at the 0.5-mile range, on sunny days, during the early sessions. The target was found with binoculars in 3.0 minutes and with unaided vision in 2.5 minutes. As one scans down the table, one notices that, more often than not, unaided vision required shorter search.

TABLE 3.2
Search Times With Binoculars and
With Unaided Vision for Comparable Conditions (One Subject)

Weather	Time of Day	Size of Target	Distance (miles)	Search Time[a] Binoculars	Unaided
Sunny	Early	Large	0.5	3.0	2.5
			1.0	6.0	6.0
			4.0	10.5	11.5
		Small	0.5	4.5	2.0
			1.0	9.0	8.5
			4.0	12.0	14.0
	Late	Large	0.5	2.0	1.0
			1.0	4.0	4.5
			4.0	9.5	9.0
		Small	0.5	3.0	1.5
			1.0	6.5	b
			4.0	8.0	9.5
Stormy	Early	Large	0.5	7.0	5.0
			1.0	10.0	9.5
			4.0	10.0	10.0
		Small	0.5	9.5	6.0
			1.0	—c	13.0
			4.0	—	—
	Late	Large	0.5	6.0	4.0
			1.0	11.5	10.0
			4.0	—	14.5
		Small	0.5	9.0	7.0
			1.0	—	12.5
			4.0	d	—
Number of comparisons on which time was shorter				4	15

[a]Times are given to the nearest one-half minute.
[b]Sighted a false target in 2.5 minutes; probably a basking shark.
[c]The mark "—" means that the target was not sighted within 15 minutes.
[d]Wasn't able to search—seasick.

In fact, in 15 of the comparisons there was more rapid spotting with the unaided eye and in only 4 with binoculars. The few conditions favoring binocular search involved sunny weather and a far range. Even this was consistent only for the small target.

Some incidentals may be noted. As would be expected, search was somewhat more successful in sunny than in stormy weather. The slight difference favoring late search was possibly due to a glint or silhouette produced by the low sun.

A similar analysis was made of Dionne's efforts. The comparisons again favored unaided vision, but not by so wide a margin: 14 to 8. There were only two targets that she failed to find. At this point, it may be noted that although the experiment was complete with only one subject, it was very useful to have had two. The fact of the smaller difference for Dionne indicates that there was some advantage for the second half of the sessions. The targets might have been easier or the subject more experienced. If only Hawkeye had been a subject, there could be the suspicion that the entire difference could be accounted for in this way. Each subject thus served as a control for the other, even though no attempt is made to generalize the findings to other persons.

Summary of the Experiment

Two arrangements for searching were compared, the use of 7 × 50 binoculars and unaided vision. Rather than compare effectiveness during typical search operations, an artificial situation was devised in which a large number of targets was dropped, under a variety of conditions. The experiment required two sessions a day over a period of four days. It had to be conducted when weather conditions were as desired. Two subjects were tested. Each did better with unaided vision than with binoculars. For each, the performance of the other subject provided a control that showed that the difference was genuine and could not be attributed to unfair testing. Each of the subjects has a good reason for doing subsequent searching without binoculars.

IMPROVEMENT 2: MORE DATA GIVES HIGHER RELIABILITY

Suppose, once again, that the experimenters had decided to conduct their experiment entirely under real-life conditions. Here is how they might have gone about it. Whenever a rescue mission came up, one of the two subject-experimenters would use binoculars, and the other would use unaided vision. On the next mission, they would switch, and so on. Unfortunately for the experimenters (fortunately for the potential victims), such rescue missions are few and far between. If there were to be 48 search missions, they would cover not a week or two, but rather several years. It is unlikely that the situation regarding crew, commander, and vessel would be stable enough for the experiment, once started, to be completed. The experimenters would probably have to settle for only a few trials, with and without binoculars. Needless to say, in just a few trials the other variables could hardly have been depended on to average out at the same level. These other variables include size and color of the target, searcher's knowledge of the appearance of the target, visibility, height of waves, etc. Although to some extent these differences are controlled for by having one searcher use binoculars and the other not use binoculars during the same mission, it is not a very effective control. Each searcher will make an *individual* decision based on that searcher's own performance.

Not only would there be fewer missions in an experiment conducted at the pace of real life but these "trials" would also not provide as much information on the searcher's behavior. In many missions, the target is not found by a searcher. It may be located by another vessel or by an aircraft or never located. There is no way of knowing whether a searcher has failed because of poor technique or because there had been no target in the range of vision, with or without binoculars.

By use of the artificial experiment, much more of the searcher's relevant behavior—that which occurs when there is a target that could be sighted—is measured. It is thus much closer than the real-life experiment to approaching the infinite experiment. It will be recalled, that the infinite experiment would absolutely take care of unsystematic trial-to-trial variability (if only it could be performed). The improvement thus represented is therefore one of increased reliability. We would be much more confident that the result found in 48 trials would be the same as in 48 other trials than if the experiment were restricted to the few trials possible in an experiment that duplicates the real world.

The experimenters also tried to control certain identified other variables: time of day, weather, size of target, and distance of target. From the standpoint of internal validity, they would have been at least as well off using just one level of each of these other variables. However, don't forget about them; they'll be back!

Because of their ability to have targets dropped at their convenience, the experimenters were able further to improve reliability. To begin with, they could drop either size of target at will. They were not able to produce any kind of weather they wanted, as in night visual approaches, but they were able to *balance* the circumstances for the trials with and without binoculars: both weather and time of day. We have seen that the experimenters were even able to make a comparison between use and nonuse of binoculars for every subcondition employed. If the experiment had been done only during real search, such niceties would have been out of the question. It would have been hard enough getting a respectable number of trials of any sort.

AN EXPERIMENT ON CHOOSING THE RIGHT ALTIMETER

Let us consider another possible but equally farfetched situation. Charles Augustus Landbug (known to his friends as the Lone Sparrow) is finally about to buy his very own little airplane. He has settled on the Z3 Sky-rocket but still has a choice to make. There is the standard altimeter with one dial and two pointers, as is shown at the left (a) in Figure 3.4. For a

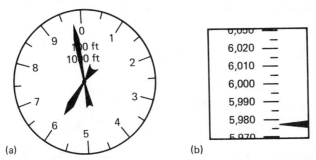

(a)   (b)

Figure 3.4 The standard two-pointer altimeter, on the left (a), and the Unitized Insto-Glance, on the right (b).

slight additional charge ($42.80), he can get the new Unitized Insto-Glance altimeter. This is the one shown to the right (b). The dial itself slides up or down when the pointer gets close to the margin. Although he worried about money in making his choice of plane, he is now more worried about his neck. He has just reread an old creased newspaper clipping he keeps in his wallet.

Ibiza, Spain (UPI) 1–7–72—An Iberia Caravelle jetliner hit the highest mountain on this Mediterranean tourist island Friday killing all 104 persons on board.

The Spanish plane carried 6 crew members and 98 passengers, including 6 babies, on a flight from Madrid and Valencia to Ibiza, the airline company said. An Iberia spokesman said there were only two foreigners aboard the plane—Jeffrey D. Dessak of New York and Dieter Fricker of Dusseldorf, Germany.

The official Spanish news agency Cifra reported the last radio contact with the plane came when the pilot reported his position over the small Balearic island of Cunillera, 12 miles from Ibiza Airport, and asked for permission to descend to 5,500 feet.

"Get me a beer ready, we are here," Ibiza Airport sources quoted Capt. Jose Luis Ballester as saying, according to Cifra.

The control tower gave Ballester, 37, a veteran of 7,000 flying hours and a father of 6 children, permission to descend.

According to the news agency, the plane must have been flying much lower than its crew thought.

"The plane did not hit the mountainside head on," the agency reported. "Its front end was not as badly smashed as the middle and rear section, which completely disintegrated. It appears the pilot saw the mountain at the last moment and tried to climb away from it." [reproduced by permission from United Press International]

Deciding What Kind of Experiment To Do

Landbug has talked with friends, of various levels of ignorance, who disagree among themselves on which altimeter is more dependable. What he would like to do is to make a quick and safe experiment. Charles' first thinking is pretty expansive. He will contact the manufacturer to find whether they have a trainer simulator. He could take the plane out over an imaginary terrain and fly imaginary courses using each of the altimeters. He could then decide which one he likes better. But then he thinks to himself, "I'll bet the pilot of that Spanish plane liked the altimeter he was using; I'd better get some better information of how well I use the two altimeters." His next thought is that there is probably a graphic recorder that would show just how he used each of his controls and what the readings were on each of the dials. However, he isn't quite sure of what he would make of all of this information. Anyway, he decides to call the Skyrocket Company and talk to them about the chances of using a simulator, which he assumes they have developed. They tell him they are flattered but that they have never gotten that kind of government money.

He then remembers that he really wouldn't have known what to do with the data he would have gotten with a simulator. Now a very simple possibility occurs to him. He makes a cardboard mockup of each dial and arranges to have a photographer take pictures of them with the pointers in all kinds of positions. These provide the basis for an experiment on himself in reading the two altimeters.

Conducting the Experiment

In all, there are 120 pictures for each dial showing a variety of altitudes, the same values being shown in both. Landbug shuffles togeth-

er the pictures for the standard dial, and separates them into two piles of
60. He calls one of the piles Set A and the other, Set D. Similarly, the
piles for the new altimeter are called Set B and Set C. These letters
indicate the order in which he is going to be tested on the readings. He is
going to need two tape recorders (or one tricky recorder) to carry out the
experiment. He is going to require himself to make a reading every 5 seconds. To do this, he makes a tape in which he looks at the sweep-second
hand of a clock and says "now" every 5 seconds. He listens to this
through earphones while he is tested. He must turn to the next picture
each time he hears "now." The second tape recorder is, of course, used
to take down his calls of altitude. Each set requires five minutes of
testing.

Analyzing the Data

The first way of analyzing the data that occurs to Charles is to list
for each dial the true altitudes and opposite each to write the altitude
called out. The error is thus found on each call, and the average amount
of error is computed for each of the altimeters. Not too much difference
was found. The average amount of error for the standard dial was 12 feet
and that for the new dial was 8 feet. A different kind of analysis is shown
in Figure 3.5. Each of the correct altitudes is shown as a mark on the
horizontal axis. Along with this is shown the error in the reading—positive (too high) or negative. In the portion shown, there is not much difference between dials, with one exception. Where the correct reading was
5,980 feet, it was read as 6,975 feet, an error of almost 1,000 feet on the
old dial. If we reexamine Figure 3.4, we can understand how this error
came about. There were several other peculiar, but not so obvious, large
errors made with the old dial. On this basis, Charles decides to pay the
extra $42.80.

Summary of Experiment

In order to decide whether to have a special altimeter installed in a
plane, Charles Landbug devised an experiment that he conducted on himself. This consisted of reading off altitudes from photographed dials at a
forced pace. For the most part, there was little difference in error. However, with the conventional altimeter he made a few very large errors.
The confusions were sometimes understandable, but sometimes not.
Charles reasons that it would be dangerous to fly using such an instru-
ment and selects the new altimeter.

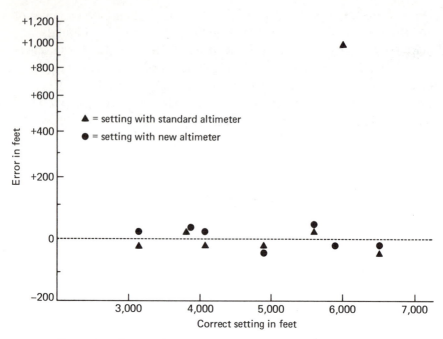

Figure 3.5   Portion of graph of errors against setting: altimeter experiment.

IMPROVEMENT 3: BETTER RELIABILITY BY REDUCTION OF UNSYSTEMATIC VARIABILITY

Suppose Landbug had been allowed to try out each of the dials in actual flight and had used that opportunity to do an experiment that duplicated real life. What would he have done? He would simply have made some flights, keeping the plane at the appropriate altitudes, with—of course— the help of the altimeter. The trouble is that it would be difficult to know how his behavior should be assessed. Possibly he could make two motion pictures simultaneously, one showing his dial and the other showing the terrain over which he was flying. If this had been done, only the grossest kind of assessment could have been made. It could be found that at times he was too close to mountain slopes or that he was changing altitude too often, wasting fuel. Perhaps the occasion would come up when, like the Spanish pilot, he would make a very large error. However, even if as many readings were made as trials in his experiment that improved real life, they would necessarily involve subjective assessments. How much too close was he? How stable should his altitude be? Further, they would be influenced by other variables: by his perception of the terrain (as in the experiment on night landing approaches), by his degree of cautiousness, and by his skill in flying at the intended altitude. Because his responses do not provide a good basis for assessment, we would have little stability, much unsystematic variability. Results in one experiment would not agree well with results in another unless the experiments were made very long.

In contrast, the artificial experiment he did perform required overt responses that were determined entirely by his reading of the position of the altimeter dial. They could be scored for errors quite precisely. The measurement of his performance in one experiment should be close to that which would be found under the same conditions in another experiment. Because of the reduction of unsystematic variability in assessing his response to the altimeter, he has achieved higher reliability than could be attained in an experiment that duplicates the real world.

It will be recalled that reliability can be improved in two ways. The sea rescue search exemplified improvement by increasing the number of trials. The altimeter study emphasizes the improvement by reducing unsystematic variability. Internal validity is strengthened by elimination of systematic confounding as well as by increased reliability; this was exemplified in the experiment on night visual approaches.

MORE THAN ONE IMPROVEMENT

Although each of the three experiments of this chapter improved internal validity mainly in one way, to some extent each brought about the other improvements as well. In the experiment on night visual approaches, not only was systematic confounding eliminated, but the pilot could also be given many trials within a short time, and more precise measures could be made of his altitude than in real flight. For sea rescue search, not only were there more trials than could be obtained in actual search missions, but there was also a reduction in the effects of unsystematic variability. First, much better measures could be obtained of responses, starting from the moment of the target drop. Second, the trials with and without binoculars were matched for weather, time of day, size of target, and its distance. In the altimeter study, not only was there reduction of unsystematic variability by making assessment of behavior more precise, but it was also possible to obtain more assessments in a shorter period of time than can be obtained in flight. Further, since only a few flights could have been arranged, there would have been the problem of confounding between the other variables (e.g., wind, terrain, and traffic) and the altimeter variable in the experiment that duplicates the real world. This possibility was completely eliminated in the artificial experiment. What has been said about these three experiments that improved on the real world is typically the case: They allow the strengthening of internal validity in each of the ways possible.

EXTERNAL VALIDITY: PROBLEMS OF APPROPRIATENESS

Some good news and some (possibly) bad news. No doubt the artificial experiments were better experiments than those which could have been done to duplicate the real world. The real world was "improved." That is the good news. Still, were these artificial experiments the *right* experiments to perform? Can we apply the results to the real world, which gave rise to the experiments in the first place? If the answer is no, that is the bad news. We will have gained internal validity at the cost of losing external validity.

We will now consider the problems of appropriateness raised in the three experiments of this chapter. You will see that sometimes there has been a satisfactory solution but not always. This examination will be made through the terms that make up the experimental hypothesis. An experimental hypothesis proposes a relation between an *independent* and a *dependent* variable. Thus we will look at the independent variable for appropriateness and also at the dependent variable. However, there is one more important term, which is often not stated but which is always implied: the level of *certain other variables*. You may recall that one of the possible experiments described for Jack Mozart lacked external validity because he would have practiced a different kind of piece than that which was stated in the experimental hypothesis. The third phase of examination then will look at appropriateness of levels of other variables.

It should be remembered in this examination that we are dealing with experiments that were done in response to practical needs. The best evidence obtainable, in each of the given circumstances, was sought. We have seen, in the previous chapter, that perfect internal validity is not possible, since an actual experiment can be neither ideal nor infinite.

Experiments were said to have *greater* internal validity as they approached either of these perfect experiments. In the same way, perfect external validity is not possible, since an actual experiment cannot be completely appropriate—containing the full circumstances to which the results are to be applied. There is, thus, more or less external validity to these artificial experiments as the various aspects of complete appropriateness are met. However, artificial experiments were used because experiments that duplicate the real world—and that therefore have greater appropriateness—were lacking in internal validity. Thus, we cannot reject an experiment that improves the real world because it is less appropriate than one that duplicates the real world. It makes more sense to find whether the *best* way has been found for improving the real world. To evaluate the external validity of these experiments, the meaningful comparison is thus with other artificial experiments.

Appropriateness of the Independent Variable

Generally the independent variable poses few problems in respect to appropriateness in the artificial experiment. All that the experimenter needs to make sure of is that the treatments are typical in some cases or feasible in others. Kraft and Elworth chose a flat terrain and one with a 3° slope because they were, in fact, typical of airports where jet aircraft land. Nonuse of binoculars, in the experiment on sea rescue search, may not be typical but it is feasible. It is one real way of searching. The two dials that were being compared in the altimeter study were those which Landbug had to choose between. If he had used a type of dial that could not be used in his plane, such as a view of a tiny plane in three-dimensional space, it would not have been an appropriate treatment. His hypothesis involved only the two dials that were feasible.

Appropriateness of the Dependent Variable

You will recall that for each treatment there is one value obtained of the dependent variable. The three components of each value are these: *first*, response to the treatment by the subject; *second*, assessment of the subject's responses by the experimenter; and *third*, the way in which the assessments are combined to obtain the final value for the treatment. To say that the dependent variable is appropriate means that each of these components is appropriate. That is the way in which we will now examine the three experiments of this chapter.

The Responses   Are the responses that the subject makes to the treatments in the experiment representative of those he makes in the wider world of application? For two of our experiments, there can be little question. The pilot "flys" the simulator in response to the visual information from "the ground" just as he does in actual life, so this aspect of the dependent variable is satisfactory for the experiment on night visual approaches. The subject in the sea rescue experiment scans the ocean just as in a real search mission. However, it is a different story for the altimeter experiment. The subject is not responding to altimeter readings in the usual way. He makes no corrections in a flight path. He is not even flying a plane. So, in a sense, he is doing less than he would do in real life. In another sense, he is doing more. He is making very specific readings by saying a number. This would occur rarely in flight, where the more typical question is whether the plane is in the right altitude band for the direction of flight and readings need only be within about 200 feet of the correct altitude. The responses of the real world have not been duplicated. Instead, there has been a *selective emphasis*. Can such a large departure from the real world be justified? Let us mull this over and return to the problem shortly.

The question can best be answered by comparison with other alternatives. Landbug had already come to the realization that he would not have known how to analyze data that he would obtain in real flight, nor would he have known what to make of data obtainable on a simulator. He devised his very artificial experiment because real flights would not provide a good basis for assessment of responses. Would the same be true with use of a simulator? The only real improvement over the real world would be better matching of terrain and weather conditions for flights using the two dials. Subjective judgment would still have to be made of whether he was flying at the correct altitudes. This behavior, in turn, would still be influenced by many factors: his perception of the terrain on a given flight, his cautiousness, and his skill.

Perhaps the most important real-life aspect in reading an altimeter dial was, in fact, captured. A good decision made was to use a forced-pace test. The pilot generally has a short time in which to make a reading. The alternative would have been to devise a speed test to find how many readings a minute could have been made. This is not a good idea, for two reasons. First, it is far less representative of the pilot's operation. Second, it would have introduced the problem of how to combine speed and errors. Despite all that has been written on that topic, there really is no good way, every method involving some dubious assumption (such as the relative value of each correct reading and cost of each error).

The Assessments   In respect to assessments also, the experiment on night visual approaches was highly appropriate. The altitudes flown by the pilot constituted the way in which the response to the visual scene, level or sloping, was assessed. The situation is not as clear-cut in sea rescue search. Does it really make a difference in real life whether a target is spotted in 7 minutes or in 7.5 minutes? Perhaps not. However, if the search vessel is following a linear path, as such vessels usually do, working its way back and forth within a long corridor, a failure to locate a target within a given period of time means that it will not be spotted at all or that the vessel would have to go more slowly and not cover as great an expanse of ocean. The measure of how long it took can thus be translated to the probability of locating a target—provided, of course, it is present. For the altimeter study, there was no real choice of assessments. For each reading presented, the subject's estimate was recorded.

Combining the Assessments   We see in the two experiments on aircraft altitude a different way of combining the separate assessments. Yet each was appropriate for the application of the experimental results to the real world. In night visual approaches, the experimental hypothesis was that the subject would make the systematic error of underestimating altitude and therefore of flying too low with the slanted terrain but not with the level terrain. The curves shown in Figure 3.3 allow the testing of this hypothesis. They have averaged groups of trials for 12 pilots. However, the same representation could be given for the performance of any one pilot on a group of trials with each treatment. Of course if there were just one trial for each treatment, the combination of assessments would mean nothing more than showing the altitude flown for each point along the approach path. If a group of trials is used, there is a simple average or mean altitude flown for each closer distance to the airport.

For the altimeter study, this method of combining data would not be appropriate. The successive readings are for random altitudes, not for points nearer and nearer the airport, so that the graph of successive assessments would not make much sense. Also, it would not solve the problem simply to find the mean altitude reading of the subject when using each dial. Suppose that with the old dial there tended to be large errors for most of the readings, but with the same number of errors that overestimated the altitude as that underestimated the altitude. The average estimated altitude would be very nearly the same as the average of the altitudes presented, despite the large errors.

This last point is an important one, since it has been misunderstood in a number of published experiments. Let us give a brief example to show how it comes about. Suppose that on four successive trials a dial presented the following altitudes: 3,200; 6,100; 1,250; and 4,640 feet. For this set of trials, the mean or average altitude presented was thus 3,200 + 6,100 + 1,250 + 4,640, all divided by 4. This comes out to be 3,797.5 feet. The subjects readings were: 3,260; 6,040; 1,300; and 4,590 feet. This average is also 3,797.5 feet. No one would interpret this to mean that the subject performed without error. However, the same mistake is made by finding the algebraic average of the errors, calling a reading that is too high a *plus*, and one that is too low a *minus*. In the preceding four trials there were the following errors: 3,260 − 3,200, or +60; 6,040 − 6,100, or −60; 1,300 − 1,250, or +50; 4,590 − 4,640, or −50. The errors on the four trials are: +60, −60, +50, −50. When these are averaged algebraically, the mean error comes out to be 0. That is how the mistake is made. Some other way of combining assessments is obviously required.

One solution would be to disregard whether the errors were plus or minus. For the following trials, the *absolute error*, as this value is called, is found by obtaining the mean of 60, 60, 50, and 50. Thus, the absolute error is 55 feet. It should be remarked that this value is also subject to some criticism. It does not distinguish, for example, between the performance just described from one in which errors are not evenly divided between being too high and too low but that are mostly or entirely in one direction. For example, successive errors of +60, +60, +50, and −50 give exactly the same absolute error of 55 feet. The solution has been to find two separate measures of error. The first is indeed that already described in which the plus and minus signs are used. It is called *algebraic error* and is useful for describing how much the subject's performance is too high (i.e., overestimating) or too low. The second measure is the *standard deviation*, which tells how variable the subject's errors are. It, of course, describes how much the subject fluctuates in his performance.

However, Landbug's experiment on the altimeter dials required still a different way of combining assessments. What Landbug was trying to avoid was the occurrence of very large errors. Consequently, the two dials were evaluated in terms of the percentage of very large errors that were made (off by 100 feet or more). It should be noted that this might not be the appropriate way of combining assessments if the problem was identified as which altimeter was best for landing under conditions of poor visibility. It is to be hoped that Landbug does not fly in that kind of weather.

A percentage measure was also appropriate for the experiment on sea rescue search. What was done was to use the times to estimate the percentage of targets that would have been located sooner with and without the binoculars in a real mission. This is exactly what is important in real life.

Appropriateness of Other Variables

The problem with the "wrong" experiment by Jack Mozart, in which waltzes were memorized instead of sonatas, was that of the wrong level of a most important other variable. What was found at the "waltz level" of the variable *type of musical piece* might not hold at the "sonata level." This is a case in which a key variable was placed at an inappropriate level. Let us now examine our three experiments to see how well the question of appropriateness of level was handled in regard to key variables and also to certain other variables of concern here.

Key Variables   In some experiments, there is obviously a single key variable of central importance, as in Jack Mozart's experiment. In other experiments, there may be several as in sea rescue search in which size of target, distance of target, kind of weather, and time of day are all of about equal importance.

In the experiment on night visual approaches, the key variable was pattern of lights. This is so because this variable alone gives slope information. Obviously it makes a difference whether all that can be seen on the ground is a few lights close together or a whole panorama, from which the pilot can select any pair. In order to make the results applicable to different airports, Kraft and Elworth used a number of representative patterns. Showing that even very capable experimenters may guess wrong, they state (p. 4): "It was expected that the addition of lights to the depth and width of the city would produce a better visual reference. However, data indicate that the larger, more complex light pattern may actually be detrimental if they tend to be misleading, as in the case of up-slope terrain." What happened, of course, was that the larger patterns were even more strongly taken to represent level ground, and so the illusion was heightened. We see here that it was a very good idea to duplicate the complexity of light patterns of real cities.

Similar care was indicated in the experiment on sea rescue search, in having good representation of the key variables: size, distance, weather, and time of day. In the altimeter study, the only control of a key variable was to keep the *range of altitudes* the same as that which would be encountered in real flight. The key variable of dial movement was not duplicated. Landbug looked at stationary settings, while in real flight the altitude is often changing. Another characteristic of real flight is that successive readings are similar to one another. A plane does not go bobbing up and down randomly. This is exactly how the successive trials were presented in the altimeter experiment—randomly. One of the departures from the real world made dial reading easier, and the other probably made it more difficult. Better representation of dial changes would require motion pictures or videotapes. A rather tedious way of doing this would be through animation—single shots with small changes between them—the method of making cartoons. The spoken word *read* could then be put on the sound track.

Simultaneous Activities   In some artificial experiments, the entire task is the one studied. Thus, a person engaged in sea rescue search has no other responsibilities at the time. This, of course, is not true of a pilot. Even during the limited time of an approach, in addition to flying at the right altitude, he must make the correct adjustments so that the aircraft is flying at the right angle and is not tipped to one side. He must be careful of his speed; he must watch out for other planes. All of these activities were duplicated by Kraft and Elworth. First, the pilot actually did "fly" the simulator in all respects, not only in regard to altitude. Second, additional tasks were given the pilot; he was required to "locate and report other traffic" (1968, p. 2).

The other pilot, Charles Augustus Landbug, did none of that. Not only was he freed from making his usual responses to altitude readings, he had no additional activities. It would have been a good idea for him to have had some secondary activity. Perhaps the new dial is better only if it can be given undivided attention. A supplemental task could have been arranged on the tape recorder that instructed him when to make a reading. For example, an instruction could have been given requiring him to sound a buzzer.

Stress   All the experiments described in this chapter have been matters of life and death. That is no mere coincidence. In addition to increasing internal validity, experiments that improve the real world very often make it a safer world for the subject. However, this raises the question of whether results obtained at one level of stress are applicable to another level. It has sometimes been suggested that subjects be hypnotized so that they believe they are in the real world rather than in the one of an experiment. However, the suggestion has never been made by people who are knowledgable about hypnosis. Let us see whether lack of stress was a serious question in the experiments of this chapter.

The usual effect of stress is to reduce intellectual control of behavior. It is difficult to see how lack of stress would have heightened the illusion produced by a sloping city in the experiment on night visual approaches. One would think that, if anything, more intellectual control would have reduced the illusion. Thus it could be said that this experiment obtained its results *despite* the unrepresentative absence of stress.

Stress also tends to interfere with learned or unusual ways of doing things more than it does with natural or regular ways. Viewing with binoculars is less natural than viewing with the naked eye. Hence, the absence of stress in the sea rescue experiment was advantageous to the use of the binoculars. Again, the experiment obtained its results in spite of the advantage given to use of binoculars.

Compression   The improvement of reliability in these artificial experiments over those which could have been done by simply duplicating the real world was in large part achieved by being able to provide trials much closer together in time. Thus, much more behavior could be measured in a convenient amount of time. The artificial world, then, is often *compressed* in comparison with the real world. How would this compression affect the external validity of the conclusions?

There was the least compression of the three experiments in that on night visual approaches. Also, the obtaining of more trials was least important in that experiment. It is known that practice will, to some extent, reduce a visual illusion. Hence, again there should have been less illusion in the experiment using the simulator than in real-life approaches. The results were obtained in spite of the opportunity for practice.

No harm seems to have been done to the conclusion that search is more effective without binoculars as a result of the compression of trials. It is true that a lookout might behave quite differently in the usual long vigilance condition than in the brisk race between searchers that characterized the sea rescue experiment. However, the more usual situation should have been harder on binoculars than was this concentrated testing. The searchers did not have to use them as continuously and thus did not suffer the combined effects of weight, eyestrain, and perceptual confusion. We could say that binoculars were inferior *even* in brief testing, which favored their use.

There is no such assurance in the altimeter experiment. It is possible that in a series of trials, 5 seconds apart, the subject can become used to reading an altimeter that would be much more difficult to read if consulted only occasionally. It is difficult to know which of the two altimeters tested would benefit more from the compression of readings. Perhaps the experiment would be improved in this respect if readings were more spread out, occurring only occasionally during another task.

A Broader View of External Validity

Although the issue of external validity has been discussed at length here, it has been done from a narrow point of view. When we have asked whether the results would apply to the real world, we have, in a more systematic sense, asked whether the variables in the experiment performed represent those in a certain kind of *completely appropriate* experiment. In the next chapter, this question again concerns the real world, but there in relation to the individuals to whom the results will be applied. However, in Chapter 5 what must be represented by the experiment is not the real world but the "world of theory." Appropriateness there will mean the degree to which the concrete operations of the experiment represent the terms used in a theory. If the translation is faulty, external validity is low.

WHAT PRICE REALISM?

The examination made has been rather one-sided. Questions of appropriateness arose mainly in regard to the altimeter experiment, the one least like the real world. What was done was to show that it was a better experiment than could have been done with a more realistic simulator, but that there were still improvements that could have been made. However, the examination may be turned around to ask whether a high level of realism in an experiment—which has made it an appropriate one—has sacrificed too much in the way of internal validity. Could an alternative experiment have been done that improved internal validity while not giving up too much appropriateness?

The question is really relevant only to the experiment on sea rescue search, since use of the simulator provided a high degree of internal validity for night visual approaches. The experiment on sea rescue search was exceedingly realistic, being conducted on shipboard and with subjects observing the actual sea. It doubtless lost something in the way of control of trial-to-trial variation. Could it have been made more effective if conducted in a laboratory, somewhat like night visual approaches?

   Motion pictures of the sea could be taken and projected on a wide-angle screen. Realism could be retained by using mechanically rocking platforms such as those which have been used to study seasickness. However, several problems arise. First, the motion picture scene does not show a panoramic view to allow the kind of scanning the person ordinarily does. The deficiency is especially marked for unaided vision. Furthermore, the situation is only confused by having the camera pan (move) over the scene. Second, the picture does not have as good a representation of depth as occurs in live viewing. This is not so much of a disadvantage for viewing through binoculars, where depth perception is confused anyway and so biases against unaided viewing. The use of slides instead of motion pictures would represent still a further departure from reality. Whatever effects ship motion and waves might produce are lost. These could possibly be rather different with and without binoculars. In summary, the original experiment does not seem to have been too bad a one to do. The experiment had to be highly realistic.
   Since we are still concerned with practical experiments in this chapter, it would be an oversight not to bring up the matter of expense. An experiment that requires a ship to sail according to an experimental plan, with a cooperative captain and crew and helicopter pilots, is costly. The arrangement for the Kraft and Elworth study, involving a simulator and computer, is probably even more costly. The few photographs required in the altimeter study and the loan of two tape recorders cost Landbug almost nothing. There is definitely a place for experiments that are somewhat lacking realism. In many cases, the price for a greater amount of realism may be too high.

SUMMARY

   In this chapter, three experiments were described that do not merely duplicate the real world of application but rather "improve" it. The experiment on night visual approaches tested the hypothesis that, with sloped terrain beyond the airport, pilots fly a course that brings them too near the ground, because of a visual illusion. If this experiment had been done in the real world, using real airports, there would have been systematic confounding of results by variables other than the independent variable. This confounding was eliminated by using a simulator, a device that imitates the real world. The experiment on sea rescue search tested the hypothesis that search would be more effective without binoculars than with them. If this experiment had been done in the real, routine world, very little data would have been obtained in any reasonable period of time. Because of this, reliability would have been low. By using make-believe targets, it was possible to conduct many trials in a few weeks, improving reliability.

The altimeter experiment tested the hypothesis that a new unitized dial would result in fewer large misreadings than the standard two-pointer dial. If this experiment had been done in real flight, the assessments of how well the dials could be read would have suffered from unsystematic variability from many sources. Again, reliability would have been poor. The task of reading photographed dials at a forced pace greatly reduced such variability. Thus, internal validity may be improved by doing experiments that are artificial in some sense. Three ways, then, were described of improving on the real world in order to improve internal validity. One was to devise an artificial world in which systematic confounding was eliminated. A second was to devise an experiment that would provide more data than could be obtained in the real world, giving better reliability. The third was to make assessments of responses with reduced unsystematic variability, as compared with those which could be made in the real world, again giving better reliability.

Since the experiments did not duplicate the real world, the question of external validity is raised. Did the experiments represent the real world well enough to be considered appropriate? Since experiments of the kind described here are done when those which duplicate the real world are too lacking in internal validity, there is no point to using real-world experiments for comparison. What was done was to compare the three experiments that were done with alternative ways that would have given good internal validity. Appropriateness was examined through the basic terms of the experiment: the independent variable, the dependent variable, and other variables. For the independent variable, appropriateness was found generally not to be a problem. All that was required was that the treatments be typical or feasible.

For the dependent variable, three questions were asked: (1) Are the responses the same as those the subject makes in the real world? (2) Are the ways in which the responses are assessed those which are of importance in the real world? and (3) Are the assessments combined in a way that reflects the concerns in the real world? Here there were found to be many pitfalls in dealing with errors in response that can have plus and minus values, such as reading an altimeter too high or too low. Often two ways of combining assessments must be used. One measure finds whether the errors are greater in one direction than in the other. The second measure finds how variable the errors are.

In the artificial experiment, there is also concern whether those other variables that are held constant are being held constant at the right level. In some experiments, there are key variables for which the level must be close to those found in the real world. Also, the experimenters should try to duplicate levels of other activities, so that they are like those which occur in the real world. They should further ask whether the absence of stress in the experiment, as compared with the real world, would weaken their conclusions. Finally, the fact that trials are given very close together—unlike the usual situation in the real world—means that the effect of that circumstance should be analyzed.

The question was raised, What price realism? What was meant was that experiments should also be examined to see whether they have stressed realism too much at the cost of internal validity. Again, alternative ways are compared in which the experiment might be done. A final point raised that the financial cost of a very realistic experiment may be so prohibitive that only an experiment with less realism is feasible.

QUESTIONS

1. Why couldn't the experiment on night visual approaches have been done in a satisfactory way by using two real airports?

2. What was the main accomplishment of the experiment on sea rescue search?

3. Give an example of an experiment conducted in the real world in which there would be excessive unsystematic variability of assessments.

4. Summarize the ways in which an experiment that improves the real world will provide greater internal validity than one that duplicates that world.

5. How does the question of external validity relate to one of the kinds of perfect experiment described in Chapter 2?

6. What is meant by saying that the appropriateness of an experiment is examined through each of the terms involved in the experimental hypothesis?

7. Give an example of an appropriate and of an inappropriate way of combining assessments to obtain values of the dependent variable.

8. Since real-life stress conditions are seldom obtainable in artificial experiments, does that necessarily mean that they cannot be appropriate?

9. Why were two experiments that differed so greatly in respect to realism as sea rescue search and choosing the right altimeter the most reasonable experiments to have done on these problems?

STATISTICAL SUPPLEMENT

FREQUENCY DISTRIBUTIONS

In the statistical supplement to Chapter 1, a bar diagram was used to represent the value of the dependent variable (mean reaction time) for each of the two treatments, the flashing of a light (A) or the sounding of a tone (B). A more complete picture of the assessments obtained for an experimental treatment is given by the frequency distribution. This is shown for Treatment B (tone), following.

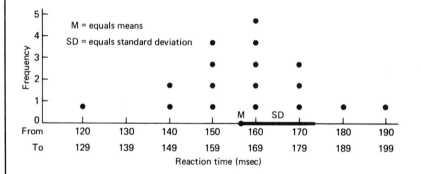

It is seen that each assessment has not been represented exactly, rather assessments are grouped in classes: Here the *class intervals* are 120–129, 130–139, 140–149, etc. The size of the present class intervals is 10 (msec).

That is the amount by which each *lower limit* increases from interval to interval (e.g., from 150 to 160 is 10 msec). The *number* of intervals here is 8; there are 8 columns. Ordinarily with more assessments than just the 17 reaction times, a somewhat larger number of class intervals would be used. For example, if there were 100 trials rather than 17, as many as 15 or even 20 intervals might be used. With 15 intervals, the lowest would be 120–124, the next lowest would be 125–129, and so on up to 190–194. In that case, the *size* of the class interval would be 5 msec.

HOW TO PREPARE A FREQUENCY DISTRIBUTION

Let us now see how the present frequency distribution was prepared. First, there were decisions on number, size, and lower and upper limits. The resulting intervals were written in a column. Then, starting with Trial 1, the different reaction times were *tallied* as to class interval. After that, the number was written. Finally, the frequency distribution already shown was drawn. The height of each column of X's represents the *frequency* of occurrence of trials falling within the class intervals. These operations are shown in the first three columns of Table 3.3.

COMPUTING THE MEAN FROM CLASS-INTERVAL DATA

In Column 4 there is shown the *midpoint* value for each class interval. Thus the midpoint of 140–149 is 144.5. We may compute the mean by a method that neglects differences within each class. First, we multiply each midpoint by the frequency within the class interval. This is shown in Column 5. Thus for the interval 170–179, the midpoint 174.5 is multiplied by 3, the frequency. ΣX is shown at the bottom of the column. When divided by N ($= 17$), the mean so computed is seen to be 163, only slightly different from the 162 value found by adding up the individual trial assessments. To be sure, there are sometimes larger discrepancies when midpoints are used rather than the individual numbers. However, the agreement is usually very good when the number of classes is 15 or more.

COMPUTING THE STANDARD DEVIATION FROM CLASS-INTERVAL DATA

Computing the standard deviation from class-interval data is done much as is the computation from individual assessments. In Column 6 there is shown the mean, which has just been computed. The value of x (i.e., $X - M_{\dot{x}}$) for the midpoint value of each interval is shown in Column 7. For example, $194.5 - 163 = +31.5$; $144.5 - 163 = -18.5$. In Column 8 each of the x values is squared. Finally, in Column 9 each of the squared values is multiplied by the frequency within the class intervals. For example with midpoint 174.5 and a frequency of 3, the product in Column 9 is 396.75. This

TABLE 3.3
Computation of the mean and standard deviation from class-interval data

1	2	3	4	5	6	7	8	9
Class Interval	Tallies	Frequencies	Midpoint	Frequency times midpoint	M_X	x	x^2	Frequency times x^2
190–199			1	194.5	194.5	163	+31.5	992.25	992.25
180–189			1	184.5	184.5	163	+21.5	462.25	462.25
170–179	|||	3	174.5	523.5	163	+11.5	132.25	396.75
160–169	++++	5	164.5	822.5	163	+ 1.5	2.25	11.25
150–159	++++	5	154.5	772.5	163	− 8.5	72.25	361.25
140–149			1	144.5	144.5	163	−18.5	342.25	342.25
130–139		0	134.5	0	163	−28.5	812.25	0
120–129			1	124.5	124.5	163	−38.5	1482.25	1482.25

$$\Sigma X_B = 2766.5 \qquad \Sigma x^2_B = 4048.25$$

and using Formula 1.1: $M_X = \dfrac{\Sigma X}{N}$ $M_B = 163$

(rounded off from 162.73529)

$$\frac{\Sigma x^2_B}{N} = 238.1$$

$$\sigma_B = 15.4 \text{ msec}$$

using Formula 2.1: $\sigma_X = \sqrt{\dfrac{\Sigma x^2_B}{N}}$

computation, again, neglects differences in assessments within any class interval. The sum of the column is seen to be 4048.25, which is Σx^2. Computing σ_x, exactly as in the statistical supplement to Chapter 2, gives a value of 15.4 msec.

It should be noted that a very straightforward method has been given here for computing the mean and standard deviation from class-interval data. That was so you would understand the principle —disregarding differences in assessments within any class. However, for strictly computational purposes, methods have been devised that are simpler and quicker. Here are a few references:

GUILFORD, J. P. , AND FRUCHTER, B., *Fundamental Statistics in Psychology and Education.* (5th ed.) New York: McGraw-Hill, 1973.

KLUGH, H. E. *Statistics, the Essentials for Research.* New York: Wiley, 1970.

WALLIS, W. A., AND ROBERTS, H. V. *Statistics, a New Approach.* New York: Free Press, 1962.

GRAPHIC REPRESENTATION OF THE MEAN AND STANDARD DEVIATION

If you will turn back to the frequency distribution at the beginning of this statistical supplement, you will notice that there is a large dot and a heavy line shown on the horizontal axis. The dot shows the position of the mean 163 msec. This is just a little to the left of the midpoint of the interval 160–169, i.e., of 164.5 msec.

The heavy line has a length of 15.9 msec, the value of the standard deviation. We see that on a frequency distribution the mean is represented by a point and the standard deviation is represented by a distance. In this particular frequency distribution, the lowest score of 122 is about 2.5 standard deviation distances below the mean score of 163. The highest score of 194 is about 2 standard deviation distances above the mean. Thus, the highest score is about 4.5 standard deviations above the lowest score. This is about typical of a frequency distribution with a small number of assessments.

PROBLEM: Compute σ_x for Treatment A from class-interval data.

Answer: 18.6

EXPERIMENTS ON REPRESENTATIVE SAMPLES

Almost all of us have experienced frustration in the supermarket. We go in with a shopping list and try to shop wisely, getting the best buys for our money. But it isn't easy. Suppose one of the items on the list is tortilla chips. The writer found arrayed on a shelf of a supermarket four different sizes of package. The 13-oz. package cost 76 cents; the 10-oz. package cost 59 cents, the 9-oz. package cost 49 cents, the 5-oz. package cost 48 cents. Now quickly!—Which was the best buy? It clearly was not the 5-oz. package, which—unbelievably—cost only one cent less than the 9-oz. package. If you have 15 or 20 items on your shopping list, it will take you a long time to find the best buys—and if you get impatient at all—you are probably going to make some mistakes.

You might say there ought to be a law. Well, there is a law. It is called the "Fair Packaging and Labeling Act." As you have seen, the consumer has not been given much help. Two psychologists set out to do an experiment to find out what *would* help (Gatewood and Perloff, 1973). About the law, they quote Cohan (1969, p. 10) as stating, "Truth in packaging . . . is one of the best non-laws on the books." In their experiment, to be described, they compared the practice on price information then current with two improvements that have been suggested: one by consumers' groups and the other by grocers.

Unlike the previous experiments described in this book, the experiment was not done on particular persons for their own guidance in the future. It was done on 75 subjects, with the intention of applying the results to all of the supermarket shoppers in the United States. With that kind of application in mind, a single-subject experiment would be out of the question. Being able to obtain results that are useful for a population rather than just for a single individual is quite an advance. Nevertheless, 75 persons do not make up much of a percentage of all shoppers. Does this *sample* represent the *population* well enough to consider the experiment as being *externally* valid? We will try to answer this question.

However, first we will find that this kind of experiment brings up some new aspects of internal validity. In the foregoing experiment, a separate group of subjects shopped with each of the three methods for giving price information. Obviously this plan has abolished sequence effects, which tormented us in Chapter 2. On the other hand, how do we know that the three groups of subjects were comparable to each other? People vary; so do groups of people.

You will find that there are quite a few ways of doing this more ambitious kind of experiment. With the exception of pointing out some really poor plans, there will not be any all-out recommendations in this chapter. The reasoning will be laid out so that you can have some guidance in making a decision in any particular case. Thus, when you finish this chapter you should be able to work out a good design for an experiment of this kind, knowing the pitfalls you have avoided and the difficulties that remain. In reading the literature, you will be able to make some good evaluations on similar efforts by other experimenters.

The areas in which you will be questioned at the end of the chapter are these:

1. The generalization permitted when many subjects are used, as compared with that permitted with the use of one subject.
2. The advantages of between-groups designs over the within-subject designs described previously.
3. The nature of between-subjects variation.
4. The logic of the different strategies for constituting groups.
5. Systematic confounding with the subject variable.
6. Reliability in between-groups designs.
7. External validity when samples are selected from a population.
8. External validity when use is made of available subjects.

AN EXPERIMENT ON METHODS OF PROVIDING PRICE INFORMATION

Here are the two methods that have been suggested so that shoppers may choose the best buy when faced with different sizes of packages. One is to employ a simple kind of circular slide rule. It is shown in Figure 4.1. The shopper turns the inner dial until the weight for a package lines up with its price, which is located on the outer dial. In one window is shown the *unit price*, the cents per ounce. (In another window is shown the ounces per dollar.) As can be seen from the setting for the 9-oz. package of tortilla chips that cost 49 cents, the unit price is less than 5-½ cents per ounce. It was, in fact, the best buy. The other method that was suggested was simply to mark the unit price on each item. Thus, the 13-, 10-, and 9-oz. packages would have been stamped 5.85, 5.90, and 5.44 cents per ounce (as well as with the total package price). Grocers admit that this obvious solution is a good one but note that it is too expensive. Gatewood

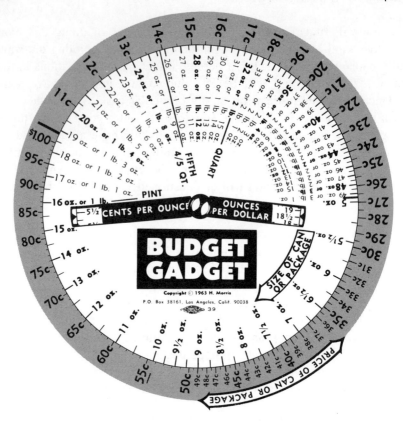

Figure 4.1 The computing device set to find price per ounce of a package of tortilla chips. The inner weight dial has been rotated so that *9 oz.*, the package weight, lines up with 49¢, the package price. In the window on the left is shown the unit price, less than 5½ cents per ounce. Reproduced by permission of H. Morris.

and Perloff (1973, p. 81) state, "Grocers, in general, have opposed this alternative. *Advertising Age* ['Grocers Moan,'1969, p. 3] has written, 'supermarket managers and suppliers complain that such a regulation will cause them to double their labor force, raise prices, or go out of business altogether.' " What they maintain is that the cheaper (maybe even paid for by the shopper) dial-a-price device is just as good.

Experimental Hypothesis

The investigators list five hypotheses that were tested. For our purposes, they may be combined in the following way: Choices may be made more accurately and more quickly with direct unit price information than with only separate size and price information (current practice) or with the help provided by the computational aid. Three methods of providing price information, then, are compared in testing this hypothesis. Treatment A is the same separate price and weight information provided in a supermarket. Treatment B is this information with added use of the computing device. Treatment C gives the price per ounce of the package

Method

Subjects "Seventy-five volunteer subjects participated in this investigation, 64 of whom were women; 60 of the 75 subjects had completed at least 1 year of college; 48 were between 20 and 29 years of age, 17 were between 30 and 39, and 10 were 40 years or older" (Gatewood and Perloff, 1973, p. 82).

They were *assigned* at random to the three treatments. Subjects in Treatment B, who were required to use the computation device, "were instructed in the use of this device and trained to a criterion of three successful price-per-unit computations" (p. 82).

Task For each product group, the packages were numbered. The subject was told to select "the most economical package" (p. 82). In all, there were nine groups of food products used. The setting was a simulated supermarket. The items, in fact, had been drawn from the shelves of a supermarket, which was a member of a chain. The subject was given an answer sheet on which to write down the number of the package that was the best buy in each of the nine product groups.

Results

Two assessments were made of the behavior of a subject on each of the nine choices: whether the choice was correct and the number of minutes taken. On the left in Table 4.1 is shown the mean number of correct choices, and the standard deviation among subjects for the three groups. It is seen that Treatment C was superior, with an average of about eight correct choices rather than the value of slightly under six for the groups having Treatments A and B. The smaller value of the standard deviation would be expected for Treatment C, since quite a few of the subjects must have made all nine choices correctly and since scores could not range above that point.

An even greater difference was found in time taken for the nine choices. This is shown in the right-hand side of the table. Treatment C

TABLE 4.1

**Correct choices and time taken
with the three kinds of price information**
**(A) separate price and weight information only; (B) added use of computing device;
(C) price per ounce information.**

	Correct Choices			Time Taken (minutes)		
Treatment	A	B	C	A	B	C
Mean	5.72	5.96	8.04	23.93	31.72	3.60
Standard Deviation	1.31	1.57	.45	10.00	9.57	1.11

required less than 4 minutes a choice on the average, while Treatment A required almost 24 minutes. Apparently working with the computing device took some additional time, with the mean for Treatment B being almost 32 minutes. Thus, it takes eight times as long to shop with the computing device than when the price per unit weight is marked.

Conclusions

The experimental hypothesis was supported. Direct unit price information did lead to more accurate choices and more rapid ones than did the other two methods. The computational device was of questionable help on accuracy and increased shopping time.

IMPROVEMENT 1 FROM USE OF MANY SUBJECTS: GENERALIZING TO A POPULATION

Suppose Gatewood and Perloff had been in a hurry and had not wanted to go to the bother of testing many subjects. The "simplest" thing they could have done would have been to test only one subject, as in experiments described previously. That might have done if they had wanted to find out whether that one subject would make better choices with unit-price information. However, they would have been rash indeed to predict that the population of supermarket shoppers would show the same differences among treatments. After all, there are a few people who are "lightning calculators." You show them two packages of tortilla chips with separate weight and price information and like a flash, they will give you the price per ounce of each—to three decimal places. Then, again, there are a few others who can't handle any kind of mechanical device. You cannot use one or a few subjects to draw the kind of inference to a population with which Gatewood and Perloff were concerned. You need a reasonably sized sample of the population. Right now we will not need to get technical about what we mean by "reasonable size" but, clearly, 25 subjects for a treatment is much better than 1 or 2 or 10.

The experiment in which many subjects are used, then, permits *generalizing* the findings to a larger population. The previous experiments were essentially those in which the generalizing was limited to a single individual—to the future behavior of the subject. Some studies are not of interest unless there can be the wider kind of generalization. In the study just described, if unit price information is to be included, it would have to be there for all shoppers. In other cases, the main advantage of the many-subject experiment is that it is economical. If a good study has been done, with sufficient subjects, it is hardly necessary to do a lot of separate studies, one on each individual of concern. Thus, it might have already been found that sea rescue search is accomplished better without binoculars for a representative sample of individuals. After that, no individual (such as Dionne or Hawkeye) would have had to do a personal experiment. Of course, one person could argue that he or she is somehow different and insist on doing the individual experiment. This brings us to the third advantage of experiments on many subjects. Because of the amount of data collected and the possibility of better experimental designs, which will be described later, they are apt to be *better* studies than those on single individuals. In summary, there are three advantages in an experiment that can be generalized to a population. First, the experiment may be of interest only in reference to a population. Second, these experiments are more economical. Third, they tend to be better experiments.

AN EXPERIMENT ON PROMPTED MENTAL PRACTICE

Here we are, back in the air again! You will recall the simulator that was used in the experiment on night visual approaches described in the last chapter. Such simulators are terribly expensive. An experimental psychologist, Dirk Prather, at the Air Force Academy, wanted a way of teaching student pilots to land a T-37 aircraft that would give the same kind of help as a simulator but without the cost (Prather, 1973). Relying entirely on real flight experience would call for more planes and instructors than would ever be available. From his study of the literature on the learning of skills, Prather came up with an idea for a most inexpensive kind of simulator: mental practice. He noted that a number of previous investigators have found that mental practice was effective for improving such motor skills as playing volleyball and shooting free throws in basketball.

Let us quote Prather on what is meant by mental practice. "Mental practice of a skill exists when the subject attempts to imagine vividly the perceptual motor actions involved in practicing the skill" (Prather, 1973, p. 353). The key idea is *mental imagery*. Prather decided to experiment on student pilots to see whether a program of mental practice that he devised would give positive results. Like Gatewood and Perloff, his interest was in applying the results to a population that was represented by the experimental subjects. However, it was not such a vast population; there are many more shoppers than there are student pilots in all training programs combined.

Experimental Hypothesis

The experimental hypothesis was that prompted mental practice, when added to the normal training program, will improve the student pilots' skill in landing a T-37 aircraft.

Method

Subjects   "The subjects were randomly selected student pilots in the undergraduate T-37 pilot training program at Williams Air Force Base. Thirteen were in the experimental group, and 10 were randomly placed in the control group. All subjects were low-experienced student pilots with approximately 20 hours in the T-41 trainer and 4 hours in the T-37" (Prather, 1973, p. 353).

Experimental Treatments   Let us first dispose of Treatment B, given to subjects in the "control group." They were given only the normal training that all past groups of subjects were given—almost nothing, except for the times in which they were being instructed during flight.

Treatment A, given to the experimental subjects, was an additional program of prompted mental practice. At four different times during the period in which pilots were trained to land, these subjects would listen through earphones to a tape recording while sitting in a mock-up (imitation) of the T-37 aircraft cockpit. The mock-up included movable throttles and a control stick. The tapes described the landing situation, at first in detail, and later on in bare outline, as the "prompts" were increasingly withdrawn. "The experimental subjects were told to imagine the situations as vividly as possible and to perform the same motor actions and eye movements that they would if they were in the actual landing pattern" (Prather, 1973, p. 354).

Procedure   Experimental subjects (Treatment A) were given prompted mental practice after the fourth, fifth, sixth, and seventh missions. Each session lasted between about 11 minutes and 15 minutes. Control subjects (Treatment B) were given only the normal training over the same period of time, including "some media presentations in the learning center" (Prather, 1973, p. 354).

"After the eighth actual flying mission, both the experimental and control subjects were rated by their own instructor pilots on their performance as to the techniques and procedures in the landing pattern on that particular mission. . . . The instructor pilots did not know which students were in which group" (Prather, 1973, p. 354).

Results

Ratings had a possible high value of 7 (best student) and a possible low value of 1 (poorest student). Separate ratings were given on procedures (how well the student knew what he was supposed to do) and on techniques (how well the procedures were executed). On procedures, Treatment A (experimental) subjects had a mean rating of 4.53 and Treatment B (control) subjects a mean of 4.26. On techniques, the respective values were 4.21 and 3.89.

Conclusions

It was concluded that the experimental hypothesis was supported. The main point is that mental practice did improve the real performance of landing the aircraft.

IMPROVEMENT 2 FROM USE OF MANY SUBJECTS: BETWEEN-GROUPS DESIGNS MAY BE USED

Some Experiments Are Made Possible

The study by Gatewood and Perloff (1973) on providing price information was not required to use a between-groups design. In the next section, we will consider whether it was a good idea to do so. Each subject could have been given all three treatments in a within-subject counterbalanced order, such as ABCCBA. However, that is not true of the experiment just described on prompted mental practice. You cannot teach the same subject how to land an aircraft by two methods and then compare the effectiveness of the methods. You *must* use a separate group of subjects for each method!

An experiment on training, then, requires a between-groups design. Thus, some experiments are possible only with the use of many subjects, since a between-groups design is required.

Some Sources of Internal Validity Are Eliminated

Suppose the Gatewood and Perloff experiment had used *within-subject* counterbalancing, with the order ABCCBA. This would first of all introduce sequence effects, which, as we know from Chapter 2, will not necessarily even out. Perhaps, after experiencing the convenience of Treatment C, unit-price information, the subject will be discouraged by having to use the computing device of Treatment B. Secondly, there would have to be different tasks for the three methods. The same group of product items could not be used for the different methods. This would require more or less complex ways of trying to equate the items in the product groups for the different methods.

To summarize, experiments with many subjects provide improvements over the single-subject experiment. First, they permit inferences about a population, not just for the subject tested. Second, they allow use of between-groups designs, which is required for some experiments and is beneficial for others.

AN EXPERIMENT ON TWO WAYS OF TEACHING SPANISH

We will now go into more detail on how experiments on representative samples are planned. This will be done through another make-believe case. It is just the kind of experiment that any of you could be devising in not too many years. You should pay particular attention to the method used for assigning subjects to groups.

In the state of California, most high-school students take two years of a foreign language, as this is required for entrance into the university.

In the town of Postgate, where there is one high school, about 100 students start the course in Spanish each year. There are two Spanish teachers, Ms. Crowthers and Mr. Ramos. Each teaches both the first and second years, with two periods of each. Moreover, the teaching of the first-year course is done in two sections on the same hours of the day, second and fifth periods; and the second-year course is taught the third and sixth periods. One day, in the faculty lounge, Mr. Ramos says to Ms. Crowthers, "Jeanie, I think we could get a lot more accomplished in the first two years and get the students a lot more involved if we gave up that old-fashioned reading, writing, and grammar in the first year and went over to a more relevant method like the new method they have been working out at Middle Central Kansas, where only the spoken language is used for the entire first year." Ms. Crowthers replies: "If I understand your very long sentence, Primo, it sounds like work. But life has been dull lately; I'm willing to give it a whirl." They decide that the new method should be tried out in comparison with the present method. So, together they convince their department chairman, who convinces the principal, who convinces the superintendent of schools, who convinces the school board (which is willing to be convinced about anything that doesn't cost more money).

Planning the Experiment

Ms. Crowthers and Mr. Ramos decide to talk over their experiment with Mr. Rogers, who has come to Postgate High as a counselor and who also gives a senior course in "personal exploration." He was a psychology major in college. He tells them that the most important thing about doing a study—which he learned from his text in experimental psychology, *Experimenting in Psychology*—is to keep complete records, preferably in a bound notebook with numbered pages and a table of contents.

They then start by writing in the notebook the description of the two methods of instruction. This is not difficult. For the old method, they simply write that they will use the procedures in the last revision of the syllabus, dated April 17, 1954. For the new method, they will use the mimeographed plan that Mr. Ramos was able to get from Middle Central Kansas. They also write down that they will stick with one method for a group of students. They work out a plan for keeping the laboratory book in a central place so that they can make comments on the progress of students and other noteworthy occurrences after each day of class. Naturally, they also keep their course records in the usual grade books. They realize that it is of great importance that they do their best with each of the methods and spend the same amount of time with the students. Their comments also show whether they are fulfilling their intention.

They now wisely ask each other how they will be able to tell how much Spanish a student has learned in two years. Finally, they come up with a surprisingly simple answer. If a student can listen to a question in Spanish and write an answer to it in Spanish, he has exhibited ability in both comprehension and writing. There is really no possibility, with the existing budgetary conditions, of teaching the course so that students can acquire proficiency in speech. And the one thing they agree on is that translation is not the same as knowing a language. Actually, they won't have to make use of this decision for almost two years but it is best to decide at the outset on this key of the study: the *criterion* of *proficiency*. This, too, is entered in the experimental notebook.

The experiment, of course, will start with students signing up for Spanish for the first time. One idea would be to assign the students at random first to each teacher, then to each method, which will now be called the *reading* method or the *speaking* method. There is one small complication; some of the students already know some Spanish, since they come from homes where the language is spoken. The experimenters are afraid that random assignment might result in too many of such students in some of the four groups, and too few in others. What they do about this is devise a *pretest*. This consists of 50 words written in Spanish, for which the student is required to write down the English equivalents. The scores for the 100 students are then put down in order from highest to lowest. In case of ties, the students are simply arranged in alphabetical order. The listing is shown in Table 4.2. Now a method must be selected to assign students to the four groups: (1) Ramos, reading; (2) Ramos, speaking; (3) Crowthers, reading; (4) Crowthers, speaking. The first four scores are put in that order; the next four in the reverse order; and so on. The group assignments are indicated in front of each name. Figure 4.2 shows the distributions for the four groups. It is evident that they are very similar. In all distributions, the bulk of students are seen to know very little Spanish vocabulary.

Conducting the Experiment

Each day that the two methods are used for teaching Spanish is a day on which the experiment is being conducted. However, it will be only after the end of two years of instruction that the important data may be

TABLE 4.2
Students making each score on
the pretest, and assignments to groups

Pretest Score	Group Assignment	Name	Pretest Score	Group Assignment	Name
23	1	Mendoza, J.	7	3	Alexander, J.
22	2	Lara, M.	7	4	D'Amboise, Y.
22	3	Zaragoza, D.	7	4	Fujimoto, V.
26	4	Olivera, E.	7	3	Haskel, R.
19	4	Diaz, G.	7	2	Jones, R. T.
19	3	Ruiz, V.	7	1	Lewis, I.
18	2	Avila, F.	7	1	Mazetti, T.
18	1	Becerra, J.	7	2	Olsen, C.
18	1	Cruz, C.	7	3	Smith, B.
18	2	Gutierrez, H.	7	4	Smith, H.
18	3	Miller, T.	7	4	Zimmerman, G.
18	4	Nunez, D.	6	3	Christensen, F.
18	4	Sandoval, R.	6	2	Daniels, K.
17	3	Arguelles, B.	6	1	Fisher, N.
17	2	Cruz, Y.	6	1	Graham, R.
17	1	Guzman, A.	6	2	Jones, R. H.
17	1	Juarez, L.	6	3	McIntosh, D.
17	2	Robinson, X.	6	4	Sanderson, L.
16	3	Jacobs, K.	6	4	Yee, P.
16	4	Moreno, S.	5	3	Cronkhite, P.
16	4	Sanchez, T.	5	2	Forrester, B.
15	3	Ingram, L.	5	1	Fujimoto, A.
15	2	Torres, S.	5	1	Porter, O.
14	1	Rose, G.	5	2	Young, I.
13	1	Hirashima, K.	4	3	Garner, T.
12	2	Lamb, C.	4	4	Metzger, P.
11	3	Cohen, P.	4	4	Ziliotto, M.
11	4	Howerton, D.	3	3	Anderson, R.
11	4	Smith, R.	3	2	Johnson, C.
10	3	Ciampi, V.	3	1	Petrini, P.
10	2	Jones, R. C.	3	1	Woods, C.
10	1	Kim, C.	2	2	Davis, H.
10	1	Valentine, Q.	2	3	Katz, N.
10	2	Wodenski, K.	2	4	Johnson, H.
9	3	Breck, H.	2	4	Martin, V.
9	4	Douglas, J.	2	3	Vaughan, T.
9	4	Gilbert, I.	1	2	Cooper, G.
9	3	Keller, L.	1	1	Ireland, N.
9	2	Kelley, F.	1	1	McCulloch, M.
9	1	Levine, T.	1	2	Meyer, L.
9	1	McDonald, S.	1	3	Schwartz, L.
8	2	Barnes, W.	1	4	Van Dyke, R.
8	3	Crowell, W.	0	4	Baker, C.
8	4	Graham, J.	0	3	Carpenter, F.
8	4	Hawkins, H.	0	2	Deutsch, H.
8	3	Kelly, R.	0	1	Kennedy, E.
8	2	O'Brien, J.	0	1	Kirk, B.
8	1	Miller, R.	0	2	Pinero, A.
8	1	Shapiro, L.	0	3	Reyes, M.
8	2	Watson, J.	0	4	Williams, G.

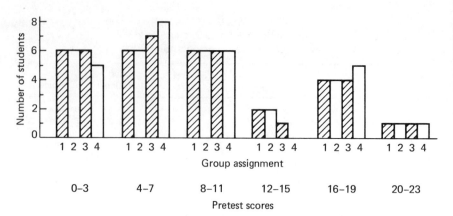

Figure 4.2 Bar graphs showing number of students assigned to each group according to score on the pretest.

obtained. During the long phase of instruction, the important aspects of the experiment are to use each method consistently and well and to note any observations that may prove to be pertinent in the evaluation of the methods. For example, the general interest of the class might be rated on each day. There also may be peculiar circumstances. A few unruly students can sometimes make it difficult to conduct a class by any method.

The important data will be collected when the class has finished, and the examination given. This should be prepared long in advance. A tape is recorded with questions in Spanish. They are asked by Ms. Crowthers of her students and by Mr. Ramos of his students, so that familiarity with the voice will not be a factor. The students take the examination in the language laboratory, where they are used to listening to tapes. After each question is asked, the student is allowed 30 seconds in which to write an answer on a slip of paper containing the question number, and the student's code number. In two sessions, the students write answers to 100 questions. So that recognition of handwriting and perhaps other idiosyncrasies of expressions will not have an influence, each teacher grades the examinations of the other. All of the answers to a given question are in one pile, with the papers from the two method groups shuffled together. Each of the teachers then grades half of the papers on each question, on a scale from 5 down to 0 (zero). They agree beforehand on the meaning of each number of points. "5" is perfect; "4" has one minor error; "3" has a bad error or two minor ones; "2" has still more errors; "1" is totally incorrect except in some aspect; "0" is totally incorrect. The highest number of points a student may obtain on the 100 questions is seen to be 500. For the sake of convenience, scores are divided by 5 and rounded to the nearest whole number. Now the highest possible total is 100. These are the criterion scores.

Analysis of Results

As it worked out, there were almost identical mean criterion scores for Ms. Crowthers' students and for Mr. Ramos' students when they received the same methods of instruction. Consequently, the results obtained are presented in the two frequency distributions of Figure 4.3, one for all students having the spoken method and one for all students having the written method. This simply shows how many students made each score, from 52 to 99. To give a coherent picture, scores are *grouped* into intervals of four; e.g., the next to the lowest interval includes persons with scores of 56, 57, 58, and 59. Just above that is the interval with scores of 60, 61, 62, and 63. Three students having the written method fell into this 60–63 points interval, and only one student having the

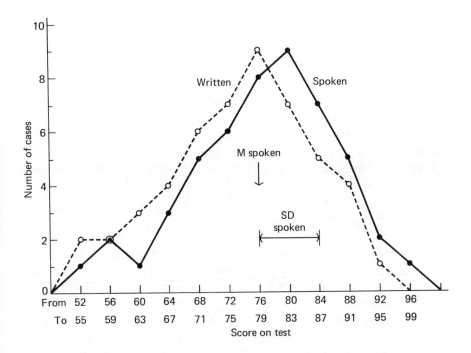

Figure 4.3 Results of the experiment on learning Spanish: frequency distributions (means and standard deviations for the spoken method).

spoken method fell into it. As can be seen, there is possibly some slight advantage favoring the spoken method; its frequency distribution is a little higher—farther to the right—than that for the written method.

However, the mean score for the spoken method was 78 and that for the written method 75, not an impressive difference. It is certainly not the kind of difference that will impress a school board enough to spend money to make a change. Of course, Crowthers and Ramos still think there is something to the new method. Maybe they just were not experienced enough at it. Still, in the hallways they are occasionally greeted with "Thanks for giving me such a good start in Spanish with that super new method. Gracias!"

ACHIEVING INTERNAL VALIDITY IN BETWEEN-GROUPS EXPERIMENTS

A New Threat to Internal Validity: Variation Between Subjects

A between-groups experiment, such as that just described, cannot be an ideal experiment for a different reason than a within-subject experiment. In an ideal experiment, the different treatments are given to the *same* subject or group of subjects at the *same* time. An actual within-subject design falls short because a subject cannot have more than one treatment at a time. An actual between-groups design falls short because it is not the same group of subjects who are being given the different treatments. Let us now look into the matter of variation between subjects and then, in the next section, consider what can be done to minimize the effects on internal validity of such variation.

Identifiable Differences

There are many differences in any group of potential subjects that are easily identified. In the experiment just described, there were boys and girls. Easy. Age is another identifiable difference. However, in identifying differences it is not necessary for subjects to fall into classes or even to have readily apparent differences. The pretest in the Spanish experiment identified subjects with different amounts of Spanish vocabulary. That is not to say that all differences that can be identified among subjects are related to the behavior that is being studied, but almost any can be. Length of hair is an identifiable difference that could be an indication of lifestyle, which in turn may affect study habits. The alphabetical position of the subject's last name may be related to ethnic background. It is hard to find a difference that could not be related somehow to the behavior in question.

The moral of all this in respect to between-groups experiments is clear. The experimenter should be on the lookout for identifiable differences. It can be seen that a study that starts with an existing difference—such as using students who sit near the front as one group and those who sit near the back as the other—is asking for trouble.

Unsystematic Variation

Give any group of individuals a test, and you will obtain a distribution of scores. Often it will have a bell-shaped appearance. This was true of students taking a pretest on Spanish vocabulary. It was true of shoppers using a computing device. It is true of rats in time required to learn a discrimination problem. For the most part, there are no previously identifiable differences among subjects to account for this variation. The differences are unsystematic.

Part of this variation could be expected from the fluctuations for each subject over time in behavior or its assessment, as described in Chapter 2. A test that is given at any one particular time will catch some of the subjects at a high point, others at a low point, etc. Further, a test does not usually measure all relevant behavior but merely samples it: luckily for some, unluckily for others. As is true for within-subject experiments, such variation in between-groups experiments is minimized by increasing the amount of behavior being assessed. In the Spanish experiment, this kind of variation was minimized in two ways. First, each student was subjected to two whole years of one of the treatments. They were thus being instructed during both favorable and unfavorable periods in time. Second, a long and comprehensive test was given. Perhaps it would have been even better to have used two or three tests over a period of a month or so, to better minimize fluctuations over time.

No matter how fluctuations within each individual are reduced there will still be variations among individuals that are not accounted for by identified differences. There are many possible ways such variations come about. The relative roles of genetic differences and basic environmental differences have been debated. However, it is the fact of variation, rather than the theories, that concerns the experimenter. Further, such differences may be particular to the experiment. It could be that most subjects would profit from the spoken method of learning Spanish but that a number of subjects would do better by the written method. Finally, there are some differences that reflect instability within the experiment. There are variations in the relations between student and teacher, between subject and experimenter.

The assessments among a group of subjects, then, will vary unsystematically, both because each individual has some instability and because individuals simply differ in their *average behavior*. The two sources of variation add together to give unsystematic variation between subjects. If two groups of subjects are each given one of the experimental treatments, there is a danger that the difference will be due to the subjects who happen to fall into each group, rather than to the treatments.

Systematic Confounding and Unreliability Revisited

In a between-groups experiment, the threat to internal validity is that the difference between the groups is not due to the treatments but to subject difference between groups. Such a subject difference could reduce or even obliterate a real treatment difference as well as falsely indicate one that does not exist. If the groups are constituted in such a way that there is a consistent identifiable difference between the two groups (e.g., boys in one group, girls in another), the independent variable has been *systematically confounded* with the subject variable. Unsystematic differences in the subjects included in the groups that occur because of limited sampling of subjects reduce the *reliability* of the values for dependent variables that are compared.

As we know, actual experiments cannot have perfect internal validity. We can evaluate procedures for constituting groups according to how closely they approach either an ideal or an infinite experiment. We have already discussed these impossible experiments in the framework of the within-subject design. For a between-groups design, an *ideal* experiment would be one in which the members of the two groups are identical. Then, treatment differences could not be due to subject differences between the treatment groups. This is impossible, because the subjects must be different individuals. However, it does suggest one approach, which is to make the groups as nearly equivalent as possible, subject for subject.

For a between-groups comparison, an *infinite* experiment would require both groups to have an infinite *number of subjects*, assuming that the method of constituting the groups avoided systematic confounding. We are here advised to use large numbers of subjects. As you will recall from the requirement for obtaining reliable averages in within-subject experiments, more trials are needed, as unsystematic fluctuations are greater. For the between-groups experiment, the comparable requirement is to use more subjects. However, we may recall that within-subject reliability is also improved by reducing variation. Since part of the variation between subjects is due to within-subject fluctuation, the previous remedy applies here too. In addition, variation will be reduced as the conditions for all subjects are made more nearly the same. If in the Spanish experiment there were differences in how the treatments were applied to the different students, we would have to expect similar variation in the progress of learning. As in within-subject experiments, precision in all aspects —independent variable, dependent variable, other variables—will improve reliability, here by reducing variability between subjects given the same treatment. As variability is decreased the same reliability may be obtained with fewer subjects.

THREE STRATEGIES FOR CONSTITUTING GROUPS, THREE BETWEEN-GROUPS DESIGNS

We will now consider three effective ways of setting up the different groups for the different treatments. They are called *between-groups designs*. We will see how each minimizes the two dangers to internal validity, systematic confounding and unreliability. These first three designs involve *assigning* subjects according to a strategy, using all available individuals. In a later section, two of the strategies will be used to devise two other between-groups designs, in which samples of subjects are *selected* from a larger population.

Random Strategy

The approach in two of the experiments was a random strategy. In the experiment on providing price information to consumers, there were 75 volunteer subjects. They were *assigned* randomly to each of the three treatments. What is meant by this is that each of the available subjects has an equal chance of being assigned to any of the treatments.

This may be done using a table of random numbers such as is shown in Table 4.3. The names of the 75 subjects are listed in alphabetical order (to simplify keeping track of them). Then a stack of cards is made, with

TABLE 4.3
A Table of
Random Digits (First 1000 Entries)[a]

Row	Column									
	1-4	5-8	9-12	13-16	17-20	21-24	25-28	29-32	33-36	37-40
1	23 15	75 48	59 01	83 72	59 93	76 24	97 08	86 95	23 03	67 44
2	05 54	55 50	43 10	53 74	35 08	90 61	18 37	44 10	96 22	13 43
3	14 87	16 03	50 32	40 43	62 23	50 05	10 03	22 11	54 38	08 34
4	38 97	67 49	51 94	05 17	58 53	78 80	59 01	94 32	42 87	16 95
5	97 31	26 17	18 99	75 53	08 70	94 25	12 58	41 54	88 21	05 13
6	11 74	26 93	81 44	33 93	08 72	32 79	73 31	18 22	64 70	68 50
7	43 36	12 88	59 11	01 64	56 23	93 00	90 04	99 43	64 07	40 36
8	93 80	62 04	78 38	26 80	44 91	55 75	11 89	32 58	47 55	25 71
9	49 54	01 31	81 08	42 98	41 87	69 53	82 96	61 77	73 80	95 27
10	36 76	87 26	33 37	94 82	15 69	41 95	96 86	70 45	27 48	38 80
11	07 09	25 23	92 24	62 71	26 07	06 55	84 53	44 67	33 84	53 20
12	43 31	00 10	81 44	86 38	03 07	52 55	51 61	48 89	74 29	46 47
13	61 57	00 63	60 06	17 36	37 75	63 14	89 51	23 35	01 74	69 93
14	31 35	28 37	99 10	77 91	89 41	31 57	97 64	48 62	58 48	69 19
15	57 04	88 65	26 27	79 59	36 82	90 52	95 65	46 35	06 53	22 54
16	09 24	34 42	00 68	72 10	71 37	30 72	97 57	56 09	29 82	76 50
17	97 95	53 50	18 40	89 48	83 29	52 23	08 25	21 22	53 26	15 87
18	93 73	25 95	70 43	78 19	88 85	56 67	16 68	26 95	99 64	45 69
19	72 62	11 12	25 00	92 26	82 64	35 66	65 94	34 71	68 75	18 67
20	61 02	07 44	18 45	37 12	07 94	95 91	73 78	66 99	53 61	93 78
21	97 83	98 54	74 33	05 59	17 18	45 47	35 41	44 22	03 42	30 00
22	89 16	09 71	92 22	23 29	06 37	35 05	54 54	89 88	43 81	63 61
23	25 96	68 82	20 62	87 17	92 65	02 82	35 28	62 84	91 95	48 83
24	81 44	33 17	19 05	04 95	48 06	74 69	00 75	67 65	01 71	65 45
25	11 32	25 49	31 42	36 23	43 86	08 62	49 76	67 42	24 52	32 45

[a]Reproduced by permission from M. G. Kendall and B. B. Smith, Randomness and random sampling numbers, *Journal of the Royal Statistical Society*, 1938, *101*, 147–166.

the name of the first person in the alphabet on the top card, and so on down. Next, a decision is made of which major column in the table to use in starting the assignments. Suppose a die is thrown, and it lands with three dots up. This means to start at the third major column of the table (Columns 9–12). Thus Aaronson (the first card) is marked with the top number in the column, 5,901. Adams (the next card) is marked 4,310, the next number down. When that column is used up, we go on to the top of the next column, until each of the 75 subjects has a random number.

The experimenter then restacks the cards in order according to the random numbers marked on them. Those 25 with the lowest numbers are put in the Treatment A pile; the next 25 in the Treatment B pile, and the 25 cards with the highest numbers in the Treatment C pile. Finally, in order to have a permanent record, the letter A, B, or C is written next to each subject's name on the alphabetical list. If you have subjects without names, such as laboratory rats, just give them names so you can make an orderly list. (Some suggested names are A7, B6, M4.)

We may see what has been accomplished by this random strategy through some contrasting examples. The investigators might have assigned the first 25 volunteers to Treatment A, the second 25 to B, and the third 25 to C. There is an obvious identifiable difference between these three groups of subjects, namely, the *order in which they appeared for the experiment*. Thus, in this bad plan there would have been systematic confounding of the independent variable method of providing price information with a subject variable. There can be no such systematic effect with the random strategy described.

However, the random strategy will not guard against inequality of groups very well unless the number of subjects in each group is sufficient. Suppose only 15 subjects were used, divided into three groups. The accidents of random placement might result in the five very best shoppers being given Treatment A. Consequently, if only 15 subjects had been used in the price information experiment, we would not be surprised if a second experiment with 15 subjects gave quite different results. Because of lack of numbers, the experiment would be unreliable. The experiment with 75 subjects, 25 to a group, is far more reliable. An experiment with 300 subjects would give even more dependable, i.e., reliable, results. Basic to the random strategy, then, is a sufficient number of subjects in each experimental group. This first between-groups experimental design is called *randomly assigned groups*.

Matching Strategy

This approach is to find an identifiable and graded difference among subjects that is thought to be related to the behavior to be assessed. Matched subjects are then placed into each of the groups. An extensive example of this has already been given for the Spanish experiment. The identifiable difference was provided by the pretest in Spanish vocabulary. This second between-groups experimental design is called *matched groups*.

There would have been no problem in using the random strategy in the Spanish experiment. Which is better? Both are equally good at preventing systematic confounding with the subject variable. In neither case is there an identifiable subject difference that is systematically different for the treatment groups. The matching strategy will equate groups better with the same number of subjects if the basis of matching (here the pretest) is really related to the behavior assessed (here the learning of Spanish). If the experimenter has guessed wrong and selected a basis for matching that is unrelated to the behavior assessed in the experiment, no harm has been done. It is as if the groups have been constituted randomly, so the equality is no worse than if a random strategy had, in fact, been used. The danger is that the experimenter will rely too heavily on matching, using a small number of subjects, only to find that the matching variable is *not* closely related to the behavior studied.

However, the difference between these two designs in respect to internal validity is minor as compared with their superiority over another design; use of *existing groups*. An example would be if the two Spanish teachers had decided to do their experiment by using the class in one high school for the spoken method and the class in another high school for the written method. This starts in with an identifiable difference among students: their high school. We cannot assume that the students in the two high schools will be equal in their ability to learn Spanish. They have come from different families, different living conditions, different earlier school instruction, etc., not to mention different teachers. Still, even if the same teacher taught in the two different schools, there would clearly be systematic confounding of the independent variable with the subject variable.

Stratified Random Strategy

This strategy might be called *mixed*, since it combines the logic of matching and that of randomness. In the Spanish experiment, the easily identified subject difference, sex, might have been used to set up two *strata*, girls and boys. There is quite a bit of evidence that girls are more advanced in language skills, on the average, than boys, so the goal would be to have the same proportion of girls and boys being given each treatment. If there were 56 girls among the 100 students, 28 would be assigned to the spoken method and 28 to the written method. The 44 boys would likewise be divided into two equal groups.

The randomness comes about in making assignments *within* each stratum, here girls or boys. Thus for the 56 girls the method of random assignment described for the experiment on price information would be used, and similarly for the 44 boys. This third effective between-groups experimental design is called *stratified randomly assigned groups*.

If the difference between the strata is related to the behavior to be assessed, then stratification provides an advantage over the *simple* random strategy. Fewer subjects are needed to obtain the same amount of reliability. If the difference between strata is unrelated to the assessed behavior, it is as if the simple random strategy had been followed. Since using strata is a kind of matching, we would expect this to be so, from the discussion of the previous section.

EXTERNAL VALIDITY: REPRESENTATION OF THE POPULATION OF INTEREST

Each of the experiments described was useful to the extent that the results could be applied to a population of interest: all supermarket shoppers, all trainees who will be instructed in landing the T-37 aircraft, all students who will be taught Spanish at Postgate High School. We know that a sample cannot be a *perfect* representative of the population because that would require one of the impossible perfect experiments: In the *completely appropriate experiment*, the subjects would, in fact, be the population to which the results were to be applied. External validity, in respect to individuals, depends on how close the method of selecting subjects for the experiment comes to that impossible goal.

This particular topic could be discussed through experiments using the same group of subjects for the different treatments, e.g., giving each subject the ABBA order of treatments. However, we will continue to use examples of experiments using between-groups designs so as better to relate problems of internal and external validity.

Selection from Populations

Very few total populations exist. There are future shoppers, pilots, and Spanish students still in their childhood or as yet unborn. In some cases, there is a large *existing* population or subpopulation from which a sample of subjects might be drawn, such as all students in the United States starting first-year Spanish. Of the three studies described in this chapter, the one that comes closest to using a subpopulation is that on mental practice in landing an aircraft. There were many more trainees at Williams Air Force Base than could be used as experimental subjects. The random strategy was followed to *select* the treatment groups from a population rather than to assign all available individuals.

Randomly Selected Groups

A technique that would give the kind of selection reported by Prather, but with an equal number of subjects in each treatment group, may now be described. It is the fourth effective between-groups design in this chapter, *randomly selected groups*.

Suppose there is a population of 810 trainees from which a total sample of 30 is to be drawn, 15 in each treatment group. Here is how to proceed. First, list the 810 names alphabetically and give them sequential numbers from 1 to 810. Now throw a die to tell on which major column to start in Table 4.3 for selecting subjects. If it comes up 5, start at the top of Column 5 (17–20), using only the last three digits of each four digit number. The first number in the column is 5,993, and the last three digits make the number 993. That tells you to select Individual No. 993 for the experiment. However, there is no such individual, there being only 810 in all. So you just go on to the next number down, which is 3,508. That means that Individual 508, or No. 508 in the alphabetical list is selected. The next individual selected is No. 223 in the alphabetical list, etc. This is continued until 30 of the 810 individuals are selected for the experiment.

If you encounter a duplicate number, such as another 508, you just skip it and go on to the next number. By simply putting individuals selected alternately in the two subject groups, there will be 15 selected for each treatment. The best way of keeping track of how many are selected is to write the number *1* to the right of the first name selected, *2* to the right of the second name selected, until the number *30* is written to the right of one of the names.

Evaluation of External Validity   This procedure has provided a random sample of 30 out of a population of 810. Whether it is a sufficiently large sample depends on the variability of the population of trainees in how well they can learn to land a T-37 aircraft. Since the trainees were obviously screened before being admitted to the training program, there is a reason to believe that the group is fairly homogeneous, with no very bad performers. Hence, even such a small sample as 30 might suffice. We expect the results, then, to approach those which would be found in a completely appropriate experiment, where (uselessly) all individuals would be given both treatments. The sample may be taken as a good representative of the population, which is to say that this aspect of external validity has been satisfied.

It still is possible that the study might not be externally valid in a larger context. If the results are to be applied to student pilots in future years, the question could be raised of whether the 810 students who make up the subpopulation are representative of the population of students over the next several years. This cannot be known positively and would require comparison of aptitude test scores, entrance requirements, etc. If these factors change greatly, the experiment would have been an *inappropriate* one in terms of subjects. The subject variable can be seen to be one of the "other" variables requiring an appropriate level. The case is entirely comparable to the "wrong" experiment by Jack Mozart, in which he attempted to apply the results of an experiment on waltzes to future memorizing of sonatas. There, type of music was the other variable that was kept at the wrong level; here, it would be the subject variable.

Relation between External and Internal Validity in Selection from a Population   *Between-groups design.* In the design used in the study of mental practice, with different treatments for different groups, there is a relation between external and internal validity. If not only the total sample, but also the sample given each treatment, is sufficiently large to represent the population, then the groups must be well equated. They could hardly be equivalent to the population without being equivalent to each other. Hence, for this kind of experiment, external validity in regard to subjects ensures an important aspect of internal validity. In between-groups experiments, the primary source of internal invalidity is difference among subjects in the treatment groups.

On the contrary, it would be possible in an experiment to obtain treatments for which subjects were well equated—contributing to internal validity—without having these subjects be a good sample of the population of interest. Suppose the first 30 trainees to arrive at the class were chosen for the experiment. They could be *assigned* to the two treatments by any of the methods previously described, including random assignment, but that would not make them representative of the population of student pilots. We thus see that equality of groups on the subject variable does not give information, one way or the other, on representativeness of a population. While external validity ensures internal validity, the reverse relation does not hold.

Within-subject design. In the experiment on price information, a within-subject design could have been used, each subject being given all three treatments. If this had been done, there would be absolutely no relation between external and internal validity. If the sample of subjects were a good representation of the population, that would make for good external validity. However, in this design invalidity comes about largely by sequence effects, task differences, and changes over time. These factors are in no way affected for better or worse by the representativeness of the sample. Similarly, effective control over these factors does nothing to enhance representativeness of the sample, i.e., external validity.

Stratified Randomly Selected Groups

Stratification can be used together with random selection within strata, like the method described for stratified random assignment of an entire group of subjects. This provides the fifth, and last, effective between-groups experimental design: *stratified randomly selected groups*. In the experiment on mental practice, test scores on spatial ability could be used as the basis of stratification. Thus, three strata of 270 trainees each could be set up in the population of 810 trainees corresponding to high, medium, and low test scores. From each stratum of 270, 5 subjects could be selected for each of the two treatments, again giving a total of 30 subjects.

If the basis of stratification (here test score on spatial ability) is related to the behavior being assessed (here learning to land a T-37 aircraft), then the stratification has been of advantage. A better representation has been made of the population with the same number of subjects, so external validity is improved. Also, since a between-groups design was employed, internal validity has similarly been aided.

Use of Available Subjects

There will always be serious questions on how well a sample represents a population when available groups, either "captive" or volunteer, are used rather than random samples of a population. This issue was already introduced when the question was brought up of whether one year's class of student pilots was a good representation of future classes who would be trained on the aircraft. This is the situation of the fictitious experiment on the teaching of Spanish. Does the class of 100 students represent future classes that will be taught? Often the situation is sufficiently permanent that the problem is not serious.

However, what are we to say about the use of 75 volunteers as representing the population of supermarket shoppers? There is no reason to believe that this is a representative sample of all shoppers. Perhaps the best argument in this particular experiment is that the sample used far exceeded the general population in educational background. Hence, if even these somewhat superior subjects could not manage the computing device effectively (Treatment B), there is not much hope that among the population of supermarket shoppers, by and large, the device will be very effective. When samples are used that are clearly not representative, it is a good idea to make a direct statement of such assumptions.

A Compilation of the Between-Groups Designs

So many different designs have been presented to you in this chapter that you must be having trouble keeping them all straight in your mind. Table 4.4 comes to your rescue! It presents a classification of the designs according to the strategy for constituting the treatment groups and the use made of potential subjects. There is also a rating of each of the six designs according to how well it solves the problems of internal and external validity. There are two cells with no entries because they represent impossible combinations of strategy and use made of potential subjects. Both matching and use of existing groups are meaningful only as ways of assigning subjects, not of selecting them.

TABLE 4.4

**Summary and Evaluation of Between-
Groups Designs for Representing a Population**

Strategy for Constituting Groups	Validity	Use Made of Potential Subjects	
		Selection	Assignment
Random	Internal	Good	Good
	External	Good	Depends on other information
Matching	Internal	—	Very good
	External	—	Depends on other information
Stratified Random	Internal	Very good	Very good
	External	Very good	Depends on other information
Existing Groups	Internal	—	Poor
	External	—	Poor

You see that the stratified-random and matching strategies are indicated as having better internal validity than the random strategy. That is because better equalization of groups is often obtained with the same number of subjects. We are never confident that existing groups are comparable because of the many possibilities for confounding. The stratified-random strategy also tends to give better external validity than the random strategy when subjects are selected from a population. That is because the population is often represented better.

We cannot make a blanket statement on external validity when assignment methods are used except in the case where existing groups are used; a major worry with existing groups is whether they do represent the same population. When subjects of a single group are assigned to the different treatments, the answer to the question of external validity depends on other information that we have about the subjects. We would like to know whether relevant dimensions such as age, schooling, and socioeconomic background are the same as for the population. In *every* evaluation made here, it has been assumed that a reasonably large number of subjects are used. If the number of subjects is small, we can have only little confidence that the population is well represented (external validity), and the equating of groups becomes more difficult (internal validity).

SUMMARY

Three new experiments were described in this chapter, all using a number of subjects in contrast with the single-subject experiments described previously. The first experiment was on methods of providing price information to consumers. In it the hypothesis was tested that marking items by price per unit of weight would result in more accurate and quicker choices of best buys than would the other two methods tested, which were (1) the current method of giving separate information of price and weight of package and (2) the use of a computing device in addition to this information. The hypothesis was supported. This kind of experiment is an advance beyond the single-subject experiment in that a generalization may be made from the sample of subjects tested to a larger population, instead of only to the future behavior of the subject.

The second experiment was used to see whether prompted mental practice helped in training pilots to land a particular aircraft. It was found the pilots given this training did perform better than those who were not given this practice. A second improvement has been seen to result from the many-subjects experiment. A between-groups experimental design becomes possible. In this kind of design, each group of subjects is given a different treatment. Experiments on methods of training, such as that described, require such designs. A within-subject design, which gives each subject all of the treatments, is impossible. The same individual cannot be taught the same skill in two different ways. Even where a within-subject design is possible, as in the experiment on price information, there are advantages to between-groups designs. Some sources of internal invalidity are eliminated, notably sequence effects and task differences.

A fictitious experiment was described in detail: two ways of teaching Spanish. A problem of special concern was that of obtaining two groups of subjects who were equal in being able to learn Spanish. Thus, it is seen that the between-groups experiment, while avoiding some sources of internal invalidity, introduces one. This is the variation between subjects. Two kinds of differences were described. First, there are identifiable differences, such as sex or age or school attended. This also includes differences that are revealed by tests, such as knowledge of Spanish vocabulary. In addition, there is unsystematic variation between subjects, which is the sum of the variation within each individual and the variation between individuals in their "average level." The experimental designs of the three experiments all avoided systematic confounding of the independent variable and the subject variable. However, when the groups that are being given the different treatments have an identified difference, such as being in different schools, confounding does occur. Reliability is increased as more subjects are used and as unsystematic variation is reduced. The latter is accomplished by making sure each subject in a group has the same conditions and by reducing the variation within each subject, as was discussed in Chapter 2.

There are three strategies for obtaining groups of subjects that are equal in respect to the behavior being assessed: random, matching, and stratified random. They were illustrated in the context of three between-groups experimental designs in which all available subjects are *assigned* to one experimental treatment group or another. In the randomly assigned groups design, each subject has an equal chance of being assigned to any group. In the matched-groups design, subjects are first ranked according to some measure related to the behavior being studied in which there are graded scores. Then matched subjects are put into the different groups. In the stratified randomly assigned groups design, subjects are first classed on some identifiable difference related to the behavior being studied. Then random assignments to the experimental groups are made within each class or stratum. All of these methods avoid systematic confounding by the subject variable. Reliability is improved as the number of subjects is increased. If the basis for matching or stratification is actually related to the behavior studied, the matching strategy and the stratified random strategy will give better reliability than will the simple random strategy, if the number of subjects is equal.

In order for an experiment to be applicable to the population of interest, the sample of subjects must be a good representation of that population. The impossible experiment that is *completely appropriate* in respect to subjects would use the entire population. External validity of an experiment can be described by how closely the selection of subjects comes to this goal. A sufficiently large random sample from a whole population would be a good representative. However, the conditions for such sampling rarely occur; very often the whole population does not even exist at any one time. The closest approach in the three studies described was that on mental practice. Random sampling was employed, using a population of trainees. Since the experiment had a between-groups design and there were two treatments, two random samples were drawn. The random approach here is that of *selection* rather than of assignment as when all available individuals must be used as subjects. The design is called *randomly selected groups.*

Although not very many individuals were selected as subjects in the mental practice experiment, the samples might well be representative of the subpopulation drawn from if there were no great amount of variation between individuals. If the intention of the experiment were to apply the results beyond the subpopulation, to classes in the future, there would be a more serious issue of external validity. The question is whether the "level" of the subject variable for the population actually sampled is appropriate for the population of application.

The relation between external and internal validity was described for experiments that select subjects from a population. In the between-groups design, if the selection has given good external validity in respect to the subject variable, this will ensure equated groups or good internal validity. However, there can be well-equated groups that do not represent the population of interest. So, good internal validity here does not ensure good external validity. For the within-subject design, there is no relation between external and internal validity, since they are determined by different factors.

Stratification can be used in conjunction with random selection just as with random assignment. This provides the fifth effective between-groups experimental design, *stratified randomly selected groups.* The population is first divided into the classes or strata, and then random selection is employed within each stratum. The same benefits occur as with random assignment if the basis for stratification is related to the behavior being studied.

When available subjects are used, either from a "captive" group or from volunteers, there is always danger of an unrepresentative sample. The experimenter must examine the situation to judge the representativeness of the sample and to foresee the consequences if it is not representative of the population of interest.

QUESTIONS

1. How do the generalizations drawn from the experiments in this chapter differ from those drawn from the experiments of the earlier chapters?

2. When is it essential to use a between-groups design rather than a within-subject design?

3. What sources of internal invalidity are eliminated by a between-groups design?

4. What are the components of between-subjects variation?

5. What is wrong when an experiment is conducted using different existing groups, such as Spanish classes at two different schools?

6. How may reliability be improved in between-groups experiments?

7. Distinguish between the two following between-groups experimental designs: randomly assigned groups and randomly selected groups.

8. What is the relation between external validity and internal validity in a between-groups design?

9. Give an example of an experiment that does provide good representativeness of one population but not of another.

STATISTICAL SUPPLEMENT

STRENGTH OF ASSOCIATION BETWEEN THE INDEPENDENT AND DEPENDENT VARIABLES

How much difference does it make whether Treatment A was given or Treatment B? The only kind of answer we have been able to give, up to now, was in terms of units of the dependent variable. Thus, for example, we have been able to say:

1. With use of ear defenders, Subject D missed 763 picks per hour; without them she missed 908. This is a difference of 145 picks per hour.
2. With a light signal, a subject had a mean reaction time of 185 msec; with a tone signal, the value was 162 msec. This is a difference of 23 msec.
3. With prompted mental practice, subjects had a mean rating of 4.21; "control" subjects had a mean rating of 3.89. This is a difference of .32 rating-scale points.

In which experiment was there the greatest effect of the inde
pendent variable: How do you compare 145 picks per hour, 23
msec, and .32 rating-scale points? You don't. One possibility of
comparable measures is to use ratios instead of differences. Thus,
for these three experiments, the ratios for the higher to lower
treatments were

$$\frac{908}{763} = 1.19, \quad \frac{185}{162} = 1.14, \quad \frac{421}{389} = 1.08$$

These are seen to be almost identical ratios. They do, in fact, tell us
something about the strength of association, but they are inadequate
for two reasons. First, this approach cannot be applied to experiments with more than two treatments such as that on methods of
providing price information. Second—and more important—is that
the ratio generally does not tell us what we want to know. That is
how *distinctly separated* are the assessments for trials or subjects
with one treatment from those with other treatments.

USE OF FREQUENCY DISTRIBUTIONS

We can put the frequency distributions for the different treatments on the same graph to see how distinctly separated they are.
Rather than use heights of columns, it makes comparisons a little
clearer to draw a line connecting the tops of the columns. (This is
called a *frequency polygon.*) This method has already been shown
in Figure 4.3 for the experiment on learning Spanish.

We can judge from the great overlap of distributions that the
test scores for those two treatments, written and spoken, were not
distinctly different.

Now let us look at the distributions for the reaction-time
experiment described in the statistical supplement to Chapter 1.
This is shown in Figure 4.4. Remember that these are make-believe
data. We will now make believe that they were obtained in a
between-groups experiment. Each of the reaction times thus represents the mean for one of the subjects, 17 subjects having been
given each treatment. This example could be just as well developed
in terms of the *within subject* experiment originally described. The
change to between-groups design is being made only to tie the
analysis to the orientation of this chapter. There seems to be a more
distinct separation of subjects—i.e., *less overlap* between distributions—for this experiment than for that on Spanish. It would be nice

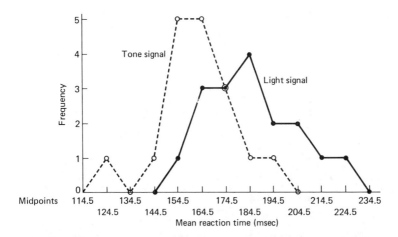

Figure 4.4 Frequency distributions for mean reaction times. Separate
distributions are shown for Treatment A (light) and Treatment B (tone).

to have a quantitative measure of distinctiveness of separation rather
than talk in such vague terms as "seems to be." Such a quantitative
measure would tell us how *strong* the association was between our
independent variable and our dependent variable.

COMPUTATION OF ω^2

We can obtain a numerical value of the strength of association
by computing ω^2. (ω is the Greek small letter *omega*; we call ω^2
omega squared.) Actually, ω^2 is one of those population parameters
that were described in the statistical supplement to Chapter 1. Its
use is described fully by Hays (1973).

Our computation using data from a sample of subjects then
yields an estimate of ω^2. This we shall call *est* ω^2.

Let us make a new graph of the results of the reaction-time
experiment. However, now we shall make no distinction as to
whether Treatment A (light) or Treatment B (tone) was used. As
you can see in Figure 4.4, this combined distribution is somewhat
more spread out than either of the separate distributions for light and
tone. The more the separate distributions differ from each other, the
larger will be the spread of the combined distribution.

If we were able to do an infinite experiment and obtained
the distributions shown in Figure 4.4 and 4.5, we would be able to
compute ω^2 directly from the parameter $\bar{\sigma}_X^2$ as follows:

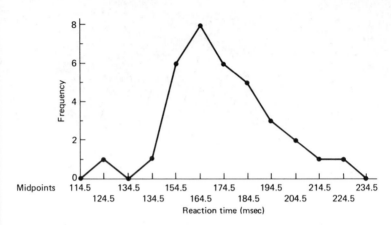

Figure 4.5   Combined frequency distribution for mean reaction times.

$$\omega^2 = \frac{\bar{\sigma}^2_{comb} - \bar{\sigma}^2_{sep}}{\bar{\sigma}^2_{comb}} \qquad \text{(Formula 4.1)}$$

However, since our data are from only a sample of the subjects implied by the infinite experiment, we must *estimate* ω^2 from the *statistic* $s^2_{X_i}$:

$$\text{est } \omega^2 = \frac{s^2_{comb} - s^2_{sep}}{s^2_{comb}} \qquad \text{(Formula 4.2)}$$

The square of the standard deviation of a distribution is often called the *variance*.

The numerator of this formula shows the difference in variance between the combined distribution and a separate distribution, either that of Treatment A or Treatment B here. By dividing this difference by the variance of the combined distribution, we put it in the form of a proportion. It answers the question: By what proportion is the variance of scores reduced in going from the combined distribution to a separate distribution?

In doing the computations, it is not necessary first to compute s_X and then to square it to get s^2_X. Remember (from Formula 2.2):

$$s_X = \sqrt{\frac{\sum x^2}{N-1}}$$

So,

$$s^2_X = \frac{\sum x^2}{N-1} \qquad \text{(Formula 4.3)}$$

In the statistical supplement to Chapter 3, we have computed Σx^2 for Treatment A (light) and Treatment B (tone).

$$s_A^2 = \frac{5894.25}{16} = 368$$

$$s_B^2 = \frac{4048.25}{16} = 253$$

Let us use those values for finding s_{sep}^2. Following Hays (p. 418), the average of s_A^2 and s_B^2 provides the estimate of s_{sep}^2, assuming the two "true" variances to be equal.

$$s_{sep}^2 = \frac{s_A^2 + s_B^2}{2}$$

(Formula 4.4)

So,

$$s_{sep}^2 = \frac{368 + 253}{2} = 310$$

When the same kind of computation is made for the combined distribution:

$$s_{comb}^2 = \frac{14188.5}{33} = 430$$

So, putting these values in Formula 4.2,

gives

$$\text{est } \omega^2 = \frac{s_{comb}^2 - s_{sep}^2}{s_{comb}^2}$$

$$\text{est } \omega^2 = \frac{430 - 310}{430}$$

$$= .28$$

This is a strong association between the independent and dependent variables. Even a value of .20 is fairly substantial. The value can never exceed 1; however, that value is rarely approached. In contrast, the computation for the experiment on Spanish is:

$$\text{est } \omega^2 = \frac{96.6 - 93.9}{96.6}$$

$$= .03$$

This is a very weak association between the independent and dependent variables.

Uses of ω^2

You will note that for s_{sep}^2 it is necessary to assume that the true variances for the two treatments are equal. In the reaction time experiment, the assumption was reasonable because the variances for the two treatments were very close in size. The same is true for the Spanish experiment. However, for the experiment on price information (Table 4.1), Treatment C had a much smaller standard deviation, especially for Time Taken, than did the other two treatments. The three variances (standard deviation squared) are seen to be 100, 92, and 1.2 for Treatments A, B, and C. With this amount of difference, there is no direct way of computing ω^2. However, it does not matter that there are three treatments rather than two. As long as the variances are close in value, a meaningful value of ω^2 may be computed for any number of treatments using for s_{sep}^2 the mean value for the treatments. Also, as was indicated, ω^2 can be computed from within-subject data just as well as from between-groups data. The variance is that among trials rather than that among subjects.

INTERPRETATION OF ω^2

We can look on ω^2 as telling us the amount by which *uncertainty* is reduced by knowledge of experimental treatment. In the reaction time experiment, we know something about an individual subject's mean score from knowledge of the experimental treatment. Still, within each treatment scores do vary. Our uncertainty is measured by the variance of scores in the distribution. If we do not know the treatment being given a subject, our uncertainty is greater —the variance of the combined distribution is larger than that for a separate treatment.

Thus, having knowledge of which treatment a subject is given *reduces* our uncertainty. As has been stated, dividing this reduction by the variance that has been reduced (s_{comb}^2), turns the answer into a proportion. Then ω^2 tells us the proportion by which uncertainty is reduced by knowledge of the experimental treatment. This is a measure of the impact on behavior of the independent variable.

PROBLEM: Compute est ω^2 for an experiment comparing Treatments C and D, with 18 subjects in each group:

$$\Sigma\, x_C^2 = 4{,}700, \qquad \Sigma\, x_D^2 = 4{,}900, \qquad \Sigma\, x_{\text{comb}}^2 = 15{,}000$$

Answer: Est. $\omega^2 = .34$.

THE ISOLATION OF AN INDEPENDENT VARIABLE

What do the following situations have in common?

1. A woman, standing inside a huge wooden box suitable for packing a piano, looks out through a peek hole at a 10-year-old Indian girl. If the girl chooses to sit on a chair near one end of the box and to press a lever over and over, after each tenth press the woman puts a marble into a metal tube so that it rolls down to a plastic tray in front of the little girl, where it lands with a "plink." If the girl just sits in the chair near the other end of the big box, where there is no lever to press, the woman sends her a marble anyway.
2. A young man sits quietly with an elastic band around his head that firmly holds in place plastic tubes fitted into his two nostrils. Although it is quite warm, he is wearing earmuffs. As he starts to inhale, he presses a button. This works two pumps that puff air scented with cloves, first into one nostril and then into the other.
3. A mother and her baby are in a room with another young woman. The mother puts the baby in a crib and says "bye-bye Tommy" and goes out the door. The baby begins to cry. Exactly 60 seconds after she has left, the mother returns to the room.

You have guessed that these are all scenes from the *avant-garde* theater of the absurd? Wrong! They are all experiments to *isolate* an independent variable. They are three of the several that will be introduced in this chapter. These experiments, done to pinpoint a variable that affects behavior, are quite different from those we have talked about before. *First*, none of them seems to provide results that can be put to immediate use, such as showing a better method for teaching Spanish or a best way of providing price information to the supermarket shopper. *Second*, the independent variable is a simpler, more *unitary* entity than those described previously, which were typically composites or "package deals." Thus, time difference at which air reaches the two nostrils is the only way in which treatments differ in the experiment using a subject with tubes in his nostrils. On the contrary, when two methods are compared for teaching Spanish, not only will they vary in spoken or written emphasis but they will also necessarily differ in the extent to which drills are used, the frequency of progress tests, order in which topics are introduced, etc., etc. *Third*, not only are the independent variables more unitary, but the treatments are more thoroughly "purified." Thus, ear-muffs were worn in the experiment on smell to reduce any disturbances. In the experiments that "improved" the real world, described in Chapter 3, the experimenter would sometimes *add* representative disturbances in order to attain external validity. In the experiment on night visual approaches, there was the simulation of other aircraft flying nearby, whose presence had to be reported.

These new experiments are done to increase our *understanding* of the variables affecting behavior. They thus form part of the science of psychology. They may or may not be immediately applicable. It certainly was not hard to find examples lacking in this kind of practicality. That explains the first characteristic noted for the three experiments listed at the beginning of this chapter. They are not necessarily "practical." The word *science* has been introduced rather reluctantly. All that is meant in regard to experimentation is the effort to understand how behavior is determined through clear and complete descriptions of the relations between independent and dependent variables.

The second characteristic, use of a unitary independent variable, indicates that a more *exacting* kind of experimental hypothesis is required if the findings are going to contribute to this understanding. When Dionne hypothesized that search would be better without binoculars than with them, she gave many reasons why this may be so: e.g., small size of field with binoculars, their elimination of depth perception, their lack of maneuverability. When the hypothesis was supported, it might have been for all of the reasons stated, for some of them, or for something she had not thought of at all. Further experiments would be required to pinpoint the effective variable or variables. In order to understand, we must *isolate* the independent variable in our hypothesis and in our experiment.

The third of the characteristics listed, use of "pure" treatments, is one of the *controls* for *contamination*. Even the ideal experiment, as previously described, is insufficient in this respect. That is, even if two treatments could be given simultaneously to the same subject, there may not be a satisfactory test of the effect of the independent variable. Both treatments could have the same impurities, the same contamination. This is easily understood if the two treatments are two different amounts of a drug given to subjects, each similarly contaminated. The effect of the pure drug might never be revealed.

Evidently the more exacting kind of experimental hypothesis is going to set a high standard for experiments. As before, this standard is embodied in one or another of the perfect experiments. An actual experiment is done to represent a perfect experiment. Here, the perfect experiment in respect to *internal validity* is a subclass of the ideal experiment that we may call the *pristine* experiment. First of all, it is an ideal experiment and all that implies; but that is only the start. In addition, this impossibly pure experiment requires that only the unitary independent variable be affected by the experimenter's manipulation, nothing else. Moreover, all other variables that might affect the behavior must be at a rigidly stable level.

One of the controls is that of "purification," which, as you have seen, could result in wearing earmuffs in August. The other controls to provide good internal validity are to eliminate *systematic confounding* with other variables. We are here considering some new aspects of confounding, not those described previously that came about because the same subject cannot be given the different treatments simultaneously. Of course, those *procedural* confoundings are a problem with these experiments, as with any experiment. On top of that, experiments to isolate a variable are subject to another kind of confounding. This kind occurs because it is impossible for the experimenter to introduce a treatment that concerns only the unitary independent variable, as demanded by the pristine experiment. Garret Hardin, a wise biologist, put the matter well in his statement: "We can never do merely one thing" (1972, p. 68). Since "the other thing" is associated with "the one thing" the experimenter wishes to do, we speak of *associative* confounding.

The first subvariety of such confounding is brought about by *artifacts* of the experiment. An experimenter would dearly love to remove only the small part of an animal's brain that figures in his experimental hypothesis. But he cannot do so. He must also damage much other tissue. If he now compares his treatment group (part of the brain removed) with his second treatment group (no operation), he has confounded the experimental variable with the *artifactual*. Is the difference in behavior between the two treatment groups due to difference in respect to possession of the part of the brain hypothesized, or is it due to general injury from the operation?

The second kind of associative confounding is not brought about by the experimental procedures as such. It is brought about by the fact that the independent variable is *"naturally"* associated with another variable. When a mother steps out of the room after leaving her baby, the crying that follows might be because the mother has gone or it might be because the baby is all alone. Getting rid of naturally associative confounding is the major task of experimenters who attempt to isolate an independent variable.

A new problem also comes up, with these new experiments, in respect to *external* validity—the question of whether the *right* experiment has been done. Experiments to isolate an independent variable are often done to test implications of a theory. Now, theories are usually very abstract, while experiments must be concrete. What is asked is that the *concrete operations* of the experiment be good representatives of the experimental hypothesis, as stated in the more abstract language of the theory. For example, it may be hypothesized that deviant behavior is more effectively reduced in adolescent boys by a rationale ("reasoning") than by punishment. The experimenter must choose some concrete form of punishment. Suppose the noise of a noxious buzzer is used (La Voie, 1973); is this a good, operational translation of the abstract term *punishment*? It is certainly not typical of punishment and perhaps has effects of a reflex nature. (It also might require the *experimenter* to wear the earmuffs.) Experiments may thus be evaluated for *operational validity*.

What this chapter should give you is a more *analytic* way of looking at experiments that have as their goal the isolation of an independent variable. Suppose you read an experimental article in which it was found that an independent variable was ineffective. You should look it over carefully to see whether the treatments were sufficiently "pure." Suppose you read an experimental article in which an independent variable was supposedly shown to have a strong effect on the dependent variable. You should ask yourself whether there was the possibility of confounding with an associated variable. If you think there is, you should be able to show it in a diagram. You might even think of a good control condition! Another thing you should look out for are experimental operations—i.e., translations of the variables of the experimental hypothesis—that are doubtful representatives of those variables. You will learn even more about isolating independent variables if you set up some experiments yourself (after searching the literature) and give them just as hard a look as you do the experiments of others.

As you read, you should prepare yourself to answer questions in the following areas:

1. The kind of experimental hypothesis tested in experiments conducted to further our understanding of behavior.
2. The purifying of experimental treatments.
3. Associative confounding and what to do about it.
4. Operational validity.
5. Choosing and using experimental subjects.
6. The reason for the sequence of topics followed in this book.

A UNITARY INDEPENDENT VARIABLE

An Experiment on the Work Ethic

Devendra Singh and William Query (1971) based an experiment on the experimental hypothesis that when a reward can be obtained either without effort or only with reasonable effort, the latter will be preferred. What we will do is to look at their quite complicated experiment, on preference for work over freeloading in children, from just this one point of view at this time. Other aspects will be brought out later in the book.

Method   Let us see how preference for one of the two treatments was obtained. First, there were the subjects: Twenty American Indian girls between 91 and 143 months of age, selected at random from girls from the second through sixth grade at the Indian School at Wahpeton, North Dakota.

They were taken from the classroom one at a time to "play a game." In the experimental room, they were shown the big box that would deliver marbles (as described at the beginning of this chapter). They were advised to get as many marbles as they could from the machine. This instruction was hardly necessary, since the girls already knew that they could either keep the marbles or else exchange them for some very attractive toys that were on display.

Before a girl showed her preference, she was taught the two different ways she could get marbles. One way was the hard way. She sat at the lever end of the box. Every tenth time she pressed the lever, a marble was delivered. She could keep this up until she got five marbles. The easy way was to sit in the chair at the other end of the box and wait for the marbles to come rolling in. In 15 seconds, she got five marbles. The stage has been set for the preference test.

On this test, the subject first pointed to the toy she wanted and then was allowed to obtain the necessary marbles. She was permitted to change chairs to sit at either end of the box as often as she wished over a period of 5 minutes. If she sat at the chair near the lever, she could, of course, press the lever to obtain a marble for each 10 presses. This is an operant conditioning schedule called FR (i.e., fixed ratio) 10. If she sat in the other chair, where there was no lever to press, she had marbles dropped at the same rate as during her previous training session using the lever. The next day she was given a similar 5-minute test session. She then either kept the marbles or got her toy.

Results   What was found was that, on the average, the girls obtained 60 percent of their marbles by pressing the lever and the other 40 percent by just sitting. Thus, not only did the subject sometimes press the lever when she could have gotten marbles simply by waiting, but also she did this more often than not. Hence the experimental hypothesis was supported: When a reward can be obtained without effort or with reasonable effort, the latter will be preferred.

Discussion   This was an experiment to increase our understanding of the nature of motivation. Neither of the two "methods" compared could be employed as such in real-world behavior, outside of the experiment. Real-life methods must be far more complicated and are concerned with periods of months and years, not of 5 minutes. One practical issue could be that of welfare payments. There are many who fear that if people can meet their needs by "freeloading" they will quit their jobs or not look for jobs. This is not the picture we obtain from the experiment. The subjects preferred to do some "honest labor" to get their rewards. That is why this is considered an experiment on the work ethic. Still, maybe, talking a social worker into recommending welfare checks might also qualify as "honest labor." The deeper issue of understanding is whether behavior must be motivated by "some underlying tissue need," at least indirectly. Thus it would be held that people have jobs in order to get money to get food in order to avoid hunger. Here the evidence points to activity that is motivated simply by a need to control an important aspect of the environment (White, 1959). Interesting, no? But now you may ask, "Why Indian girls?" You will have to wait until Chapter 8 for the answer.

The Need for a Special Situation

The same characteristic that made the experiment of little use for direct application made it a good one for testing an hypothesis concerning an isolated variable. That unitary variable was work or no work for obtaining a reward. Suppose the experiment had been done in a factory, with an eye to the possibility of immediate application. Take the job of assembling mousetraps. Couldn't we simply compare the number produced when there is a given amount of money paid for each trap assembled with the same total amount, regardless of number assembled?

We find ourselves making a decision of whether to pay the assemblers by the hour or by a piecework schedule. On the practical level, the issue has already pretty well been settled in favor of piecework. If we watch people working with the hourly "package" in effect, we will see them doing all sorts of things. Some are working quickly. Some are working slowly, to keep management's expectations low. Others are working with considerable effort but little skill, not assembling very many traps. Those who work quickly grumble about getting the same wages as the lazy and inept. Maybe, in frustration, they finally stop working so quickly. With the piecework "package" in effect, skilled workers pride themselves on how quickly they can assemble traps (while avoiding getting pinched). Those who are clumsy will probably ask to be moved to other jobs requiring less dexterity (or less tough fingers). If there is higher production under piecework—which is usually the case— it does not mean that the hypothesis is wrong. The different packages necessarily included different levels of other variables—especially social variables. The hourly wage did not allow the work ethic to show itself; the work ethic was frustrated.

PURIFYING THE TREATMENTS

An Experiment on Directional Smelling

The experimenter who inserted plastic tubes into the nostrils of a student was Georg von Békésy. He was a famous person, winner of a Nobel prize (for physiology and medicine) in 1961. He conducted this experiment (1964) at the Laboratory of Psychophysics, at Harvard University. The problem was that of finding whether time difference for a scent to reach the receptors in the two sides of the nose affects perceived direction of the source of the scent. It took an unusual person to do this experiment. First, it required somebody who would take seriously the possibility that smelling is directional. Second, it required somebody with the technique and awesome carefulness to obtain the treatments in "pure" form, with only a given time difference between nostrils. Third, and most critically, it required somebody to see that the experiment would be a contribution to scientific understanding of behavior.

You, no doubt, remember from your course in elementary psychology, that we can tell where sounds are coming from because we have two ears. If a sound source is off to the right, the sound will be a little stronger in the right ear, and, usually more importantly, it will arrive at the right ear a little earlier than at the left ear. Is it possible that we can tell the direction of a scent source in the same way? Békésy did vary the strength of the scent in the two nostrils, and he also varied the time of arrival of scent-laden air into them. We will here be concerned only with his study of time difference.

   Apparatus   Because the experiment depended on a "pure" independent variable, the apparatus was of critical importance. Figure 5.1 shows the apparatus for delivering the odor to one nostril. In the position shown, the tube to the nostril that ends in the *upper* of two Teflon plates is connected with a tube held in the lower plate, which, in turn, is connected to a tank of compressed air, with no scent. The scented air from a mixing bottle, also under pressure, just goes to a ventilator to add fragrance outside the building. When the upper plate is made to slide a short distance toward the subject, the tube to the nostril is suddenly connected to the scented air being forced from the mixing bottle. The tube containing unscented air then lines up with a hole in the upper plate and just puts a little more fresh air into the room.

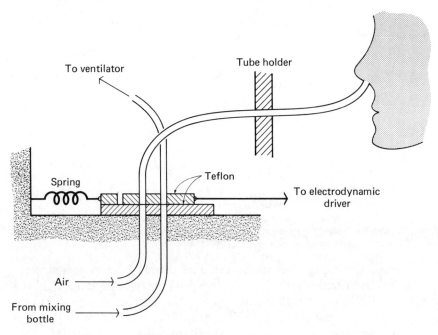

Figure 5.1   The apparatus for delivering the odor to a nostril in the experiment on directional smelling. Reproduced by permission of the *Journal of Applied Physiology* and the Nobel Foundation, from Georg von Békésy, Olfactory analogue of directional hearing, p. 370.

   Two such units were used, one for each nostril. The upper plate of each was moved by a fast, strong, electrodynamic driver. Before a trial, the experimenter adjusted a time-delay device so that he could control the difference in time when the scented air was led into each nostril. He could do this in steps of one ten-thousandth of a second (0.1 msec)!

Procedure   On any trial, the subject pressed a button as he started to inhale. This caused the scented air to be pumped into each nostril according to the time offset put in by the experimenter. The subject was ignorant of the setting, of course. He then marked a line to indicate his perception of the direction of the source of the odor, whether straight ahead or off to one side or the other.

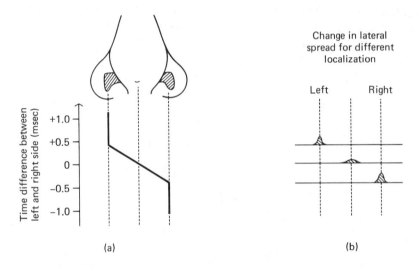

Figure 5.2   The time offsets between the stimuli to the nostrils (a) and the localizations by subjects (b) in the experiment on directional smell-ing. Reproduced by permission of the *Journal of Applied Physiology* and the Nobel Foundation, from Georg von Békésy, Olfactory an-alogue of directional hearing, p. 372.

Results   Figure 5.2 shows the time offsets (a) and judgments of subjects (b). The three levels of the independent variable are: Stimulus to left nostril 0.5 msec before stimulus to right nostril; no time difference (simultaneous); and stimulus to right nostril 0.5 msec before left nostril. As can be seen, with this very small time difference (1/2000 sec), subjects sharply localized the apparent source when the scent came into either nostril first. When the stimuli were simultaneous, they tended to localize the source as straight ahead, but with somewhat more "lateral spread."

Thus, the hypothesis was verified that difference in time of arrival of an odorous substance to the two nostrils will give the impression of direction of the source.

Discussion   Békésy notes, "It is of interest to add that the pig, an animal well known to be talented in the discovery of truffles a little below the surface of the ground, has nostril openings that are widely separated; when searching for such delicacies, it constantly moves the head from side to side in the same manner that people do when trying to locate a sound source. It is likely that the animal is making use of its capabilities of odor localization" (1967, p. 105).

However, the experiment was not done to explain why pigs could root out these expensive fungi so well. Rather, the investigator was using this demonstration to show that localization through stimuli that reach different parts of the body at slightly different times is a *general* characteristic of perception, because of the way the nervous system is organized. Localized hearing is not unique but rather is a highly developed state of this general characteristic. Thus, he was able to study "hearing" on the skin of a person's forearm instead of on the tiny organ of hearing located in the depths of the inner ear! This gave Békésy a nice big surface on which to work and an intact subject who could report his perceptions.

Extraordinary Efforts Are Sometimes Essential

To do the experiment on directional smelling properly, the treatments (time differences) had to be precise, and disturbing other variables had to be kept constant preferably *at low levels*. It is possible, as discussed previously, that *reliable* results would be obtained with less pure treatments. However, Békésy remarks, "It is interesting that as the problems of handling the stimulus were solved and the apparatus improved in precision, the performance of the observers improved likewise, and the time differences necessary for complete lateralization to one nostril became shorter" (1967, p. 103). In other words, the actual experiment had to be a good representative of the *pristine* experiment. The earmuffs were partly to prevent localization by sound cues from the drivers and pumps but mostly to keep attention focused on the sense of smell. In at least some experiments to isolate an independent variable, the independent variable must be extremely "pure" to meet the requirements of *internal validity* in testing the experimental hypothesis. This is very often true of experiments in physiological psychology. If pure drugs are not used, any effects may be obliterated. If an operation on the brain is not done especially cleanly, adjacent areas might be damaged—some that have exactly the opposite function of the area of interest to the experimenter. This, unsurprisingly, leads us to an experiment in which a part of the brain is damaged and to one in which a drug is injected.

ARTIFACTUAL CONFOUNDING

In an experiment to isolate an independent variable, systematic confounding may come about because the experiment cannot avoid introducing a second variable that might affect behavior. When this occurs, the term *associative* confounding is used. Two kinds of associative confounding may be distinguished. First, there is *artifactual* confounding, which arises from the special manipulations needed to produce one of the experimental treatments. It is discussed in the present section. Second, there is *natural* confounding, which occurs under natural conditions as well as in an experiment. This type of associative confounding is discussed in the section after this one.

Two Experiments in Physiological Psychology

Memory After Transection of the Fornix   An extremely active area of research is concerned with functions of the brain in behavior. It sometimes happens in this dangerous world that a person will suffer accidental damage to his brain and then will show either temporary or permanent disturbance of behavior. An example would be amnesia, or deficit of memory. The only real hope for gaining knowledge of the function of parts of the brain using such clinical cases would rest in examining the person's brain after he died. This does not make for very planful or feasible research. In addition, accidents are seldom precise in where they cause damage.

Consequently, controlled experimentation is done on animals. Depending on the problem, different species are used as subjects. The study of complex aspects of memory, which are of concern in human behavior, often requires the use of monkeys or even higher primates. The typical technique is to destroy a particular area of the brain, as precisely as possible, and then to compare the behavior of the animals with brain damage to the behavior of those without brain damage. After these tests, it is necessary to kill the operated animals and examine their brains to find *exactly* the nature of the damage. Needless to say, such experiments are not undertaken casually by investigators. They are planned carefully, so that the results will be useful. Also, procedures are employed to reduce distress to the subjects to the fullest extent possible.

An experiment was conducted by David Gaffan (1974, p. 1100) "to determine the part played by the hippocampus in two different kinds of memory: association and recognition." Later in this book, we will discuss how he was able to distinguish between association and recognition. Here, we will see how he made sure that whatever effects he observed were actually due to whether a structure at the base of the cerebral cortex, the hippocampus, was damaged or intact. He used rhesus monkeys as subjects.

In the ideal experiment, the same monkey would be tested at the same time, with damage to the hippocampus and without damage. This could not be done, so two groups of three monkeys each were used, one for whom the hippocampus was damaged, one for whom it was not. Supposing the groups to be well matched, the *pristine* experiment would further require that only the hippocampus be damaged for one group (no other part) and that the second group be completely undamaged. But it is impossible to damage only the hippocampus. Gaffan got as close as he could to that requirement. He selected one particular part of that structure in which to make the lesion (i.e., to damage), a narrow band of nerve fibers called the *fornix*. "Fornix transection was chosen as the form of experimental lesion of the hippocampal system since it causes the least extrahippocampal damage" (p. 1100). In other words, cutting across the fornix is the best way of restricting damage to the hippocampus, of avoiding damage to other structures.

However, it is still not possible to transect the fornix without an operation that requires use of an anesthesia and that involves considerable damage to skin, flesh, and bone if not to neighboring parts of the brain. "A 5-mm trephine hole was made in the skull and expanded with rongeurs. . . . The hemisphere on that side was gently retracted to expose the corpus callosum (the fibers connecting the two hemispheres), in which a slit was opened by aspiration [sucking through a tube] . . . the fornix was then hooked out . . . with a bent dissecting needle and transected with the aspirator" (Gaffan, 1974, p. 1102).

That is what was done to the "experimental" animals. They had the fornix cut, and a lot more. After recovery from the operation, they were to be compared on memory tasks with monkeys with an intact fornix. That is not to say that nothing at all was done to this group before the tests for memory. Gaffan operated on the second group of animals in *exactly* the same way, except that he did not hook out the fornix or transect it. The members of this latter group are hereafter called *operated control* monkeys and those of the group with the fornix removed are called *experimental* monkeys.

Diazepam-Induced Eating   Another way of learning about the function of the brain in behavior is through the use of drugs. Diazepam and similar drugs (class of benzodiazepines) have been used as mild tranquilizers in clinical practice. An observation, both of human patients and of animal subjects, is that diazepam besides calming also leads to increased eating. Understanding the function of the brain requires understanding the nature of its biochemical reactions. Finding the exact way this chemical affects behavior is a step in furthering such understanding.

Two investigators, Roy Wise and Vivien Dawson, undertook to experiment on diazepam-induced eating (1974). To study such a basic aspect of behavior (as contrasted, for example, with memory as studied by Gaffan), they were able to use laboratory rats as subjects. Again, we will delay discussion of their findings and for the present concentrate on the use of a control similar to that used in the experiment on transection of the fornix.

Unlike Gaffan, Wise and Dawson used a design in which the subject was given the two treatments. What they wished to do was compare a subject's behavior with a certain amount of diazepam in the brain and without diazepam. In the *pristine* experiment, they would do nothing else to a rat except get the diazepam into the brain. This they could not do; the diazepam had to be injected into the wall of the abdomen (an intraperitoneal injection). After 15 minutes, the diazepam would be absorbed (but not excreted), and the animal would be ready for testing.

The other treatment was to inject a neutral substance into the peritonium. Usually this substance is a physiological saline solution of the same "saltiness" as the body fluids and, thus, a solution that produces no chemical reaction. In the experiment, this is called the *placebo* treatment and the diazepam injection would be called the drug or experimental treatment. The term *placebo*, in this sense, is defined in *The American Heritage Dictionary* (1973, p. 1001) as: "An inactive substance used as a control in an experiment."

Control Conditions To Eliminate Artifactual Confounding

Suppose that Gaffan had not used operated control monkeys for comparison with the fornix-transected monkeys but had instead used monkeys with no operation at all. If that comparison had been made, there would be no way of knowing how much, if any, difference found was because of damage to the fornix or because of damage to other tissue (or, in fact, any side effects of the operation). We may show such a comparison in the following way:

	State of Fornix	
	Transected	Intact
State of Surrounding Area		
Injured	Operated group	
Uninjured		Unoperated group

The comparison is internally invalid because the two groups of monkeys differ not only in state of fornix but also in state of surrounding area. What is required is a comparison that in the diagram differs only horizontally, showing that there is the same level of the other variable. This is provided by comparing the fornix-transected group with an appropriate control group, that is to say, with a group having the appropriate control condition of differing only in respect to state of fornix. When that is done, the diagram becomes:

	State of Fornix	
	Transected	Intact
State of Surrounding Area		
Injured	Experimental operative group	Operative control group

Now the comparison is made straight across, so that confounding between state of fornix and state of surrounding area has been eliminated.

Although Wise and Dawson (1974) used a design giving each subject the two treatments rather than a between-groups design in their study of the effects of diazepam, the problem of confounding and the solution to it were logically the same as in Gaffan's experiment. If they had used noninjected animals for comparison with those injected with diazepam, the comparison would have been "vertical" in the diagram as well as "horizontal," indicating that the two treatments differed in two ways rather than in just one:

	Introduction of Diazepam	
	Yes	No
Use of an Injection		
Yes	Diazepam treatment	
No		No injection treatment

The remedy for this confounding was to change the comparison treatment so that the treatments differed only in respect to the introduction of diazepam:

	Introduction of Diazepam	
	Yes	No
Use of an Injection		
Yes	Diazepam treatment	Saline placebo treatment

Again the comparison is made straight across, so that confounding was eliminated, this time between introduction of diazepam and use of an injection.

In these experiments, there was an *active* level of the independent variable—fornix transected, diazepam introduced—and also an *inactive* level—fornix intact, no diazepam introduced. The active level is one in which the experimenter *did* something. The artifactual variable also has an active level (injury of surrounding area, injection) and an inactive level (no injury, no injection).

In both cases, part of the *active* treatment, transection of fornix or introduction of diazepam, was the production of an *active* level of another variable. This is called an *artifact* of the experiment. This association does not exist in nature. It is brought about solely by the experimenter's manipulation. If a comparison were to be made with a no-treatment group (the inactive level), there would be a difference, not only of level of independent variable but also of the level of the artifactual variable. If the pristine experiment had been possible, there would not be an artifact, because the experimenter could manipulate only the independent variable. Simply forgetting about the existence of the artifact by use of completely *inactive* comparison treatments (no operation, no injection) takes us far from an ideal experiment, let alone a pristine one. Use of *active* control conditions (partial operation, placebo) brings the experimenter close to an ideal experiment. The two treatments do not differ on the artifactual variable. The experiment is somewhat farther from the pristine experiment—there is nothing in the experimental hypotheses about destruction of other tissues or peritoneal injections—but certainly closer to it than the experiment comparing the active treatment with a completely inactive treatment. We may summarize artifactual confounding and the use of control conditions for eliminating the confounding as follows:

1. EXPERIMENT WITH ARTIFACTUAL CONFOUNDING

	Independent Variable	
	Active level	Inactive level
Artifactual Variable		
Active level	Active treatment	
Inactive level		Inactive treatment

2. EXPERIMENT WITH CONTROL FOR ARTIFACTUAL CONFOUNDING

Independent Variable

	Active level	Inactive level
Artifactual Variable		
Active level	Experimental condition	Control condition

The control condition here consists of the *inactive* level of the independent variable and the *active* level of the artifactual variable. It is compared with the experimental condition that consists of the active level of both these variables.

When a within-subject design is used, each subject is given both treatments, the experimental condition and the control condition. In a between-groups design, the two groups compared are usually called the *experimental group* and the *control group*.

NATURAL CONFOUNDING

An Experiment on Babies' Crying

Ask a mother whether her baby ever cries as a result of her leaving, and you probably will get a yes response. Researchers in social development would like a more objective answer. They would like to know just how important an attachment figure (usually the mother) is in the socialization of a child. If there really is a phase in the infant's life when it is distressed because the mother leaves, this behavior could be used in charting the progress of social development.

That is the background of an experiment on attachment behaviors in human infants that was conducted by two developmental psychologists, Don Fleener and Robert Cairns (1970). When they looked back at previous studies, they found large discrepancies. "While one researcher has estimated that the onset of an infant's discriminative crying on being separated from its mother normally occurs within 3–6 months after birth, other workers have concluded that the median age of onset of such behavior is 7–9 lunar months after birth" (p. 215).

Method   The investigators noted that the independent variable had not been manipulated as precisely in previous experiments as was implied in the experimental hypothesis. The infants *did* cry when the mother left. But maybe that would have occurred if any person they had been with for a short time had left. So they set up the experimental situation of the infant having both the mother and an experimental assistant (another young woman) present. These women then left one at a time, the infant's behavior being recorded both on motion pictures and on sound tapes. When one of them left the room, she stayed away for 60 seconds. For the different infants tested, one of four orders of treatment was used between the mother (M) and the experimental assistant (E) leaving: EMEM, EMME, MEEM, and MEME.

"Blind" ratings were made on amount of crying by judges independently working with the films and tapes, without knowledge of whether any episode being observed was during M's or E's absence. The crying score was the number of 5-second blocks during which the infant was judged as crying. The use of blind ratings is an important control for *experimenter bias*, which was discussed in Chapter 2. If the judge had the opinion that there would be more crying on departure of the mother, there would be a good chance that he or she would see it (and hear it) that way.

Results   Different age groups were used in the experiment. Not much difference in the amount of crying was found for the youngest infants. However, for infants between 12 and 14 months of age there was a clear difference. The crying score for these 15 infants was 11.67 with the mother absent and 8.27 with the experimental assistant absent. The *specific* reaction to the departure of the mother was thus found to occur much later than had been previously reported. A problem was that once the infant started to cry, it would cry when almost anything (or nothing) happened.

A Control Condition To Remove Confounding by a Wider Variable

The previous experiments were regarded as faulty by Fleener and Cairns in that the results could not show whether an infant cried because it was specifically the mother who had left, thus:

	Mother Leaving	
	Yes	No
Anyone Leaving		
Yes	M leaving	
No		M not leaving

This will be recognized as a comparison of treatments that differ in two variables. Here the *pristine* experiment is not possible on logical grounds. We cannot have the mother leave without having *someone* leave. Fleener and Cairns removed this confounding by the use of the assistant E, who *was* someone, and not M, the mother. Thus:

	Mother Leaving	
	Yes	No
Anyone Leaving		
Yes	M leaving	E leaving

These two tables are essentially the same as described for artifactual confounding. First, there was:

	Independent Variable	
	Active level	Inactive level
Wider Variable		
Active level	Active treatment	
Inactive level		Inactive treatment

Use of an active *control condition*, E leaving, allows horizontal comparison on only one variable, the independent variable of the hypothesis:

	Independent Variable	
	Active level	Inactive level
Wider Variable		
Active level	Experimental condition	Control condition

Again we see that there had been confounding when an active treatment (M's leaving) had been compared with an inactive treatment (M's not leaving). That is the active level of the independent variable also included the active level of the wider variable; the inactive level of the independent variable also included the inactive level of the wider variable. This confounding was removed by testing at the active level of the other variable—here, not an artifactual variable but a wider variable than was stated in the experimental hypothesis.

Control Conditions To Eliminate Confounding by a Secondary Variable

In a review of the literature on human attachment (1974), Leslie Jordan Cohen notes that Fleener and Cairns (1970) had avoided still another kind of confounding that earlier experiments had not. She points out that in earlier studies, the infant had been left alone when the mother departed. Hence it would be hard to say that the crying was due to the fact that the mother had gone or because the child was all alone. From that point of view, the earlier experiment may be shown as:

	Mother Leaving	
	Yes	No
Someone Staying		
No	M leaving	
Yes		M not leaving

This particular confounding was removed by having either E or M stay when the other had left. Now the experiment is shown as:

	Mother Leaving	
	Yes	No
Someone Staying		
Yes	M leaving	E leaving

Apparently all confounding has been eliminated, and the desired straight-across comparison on only the independent variable may be made. But, Cohen (1974) points out, there still has not been sufficient control. This can be seen by the following representation of the Fleener and Cairns (1970) experiment:

	Mother Leaving	
	Yes	No
Person Staying		
E	M leaving	
M		E leaving

Now the independent variable mother leaving (Yes or No) has been confounded with another variable, person staying (E or M). Cohen suggests a plan to eliminate that confounding. It would be to involve a third adult in the experiment, a constantly staying person. Now, under one treatment, the mother and the staying person would be present with the child, and then the mother would leave. Under the other treatment, the experimental assistant and the staying person would be present with the infant, and then the experimental assistant would leave.

The design for control of the staying person may be shown as:

	Mother Leaving	
	Yes	No
Person Staying		
The same third person	M leaving	E leaving

At last (we hope!) the comparison can truly be made straight across, with the only difference being in the independent variable, mother leaving.

We have just seen two cases of confounding by a *secondary* variable. That is a variable that is produced by the experimental manipulation but that is not unique to the experiment; it is typical under natural circumstances. There was first the secondary variable of someone staying. When its mother leaves, the infant is quite often left alone. Being left alone is secondary to the mother's leaving. When the mother and a friend are both with an infant, they will usually leave one at a time, not together. Depending on which one leaves, there is produced one or the other level of a secondary variable—person staying.

In most general terms for associative confounding, whether artifactual or natural (and with the natural condition, whether with a wider variable or with a secondary variable), there may be the same representation:

EXPERIMENT WITH ASSOCIATIVE CONFOUNDING

	Independent Variable	
	Active level	Inactive level
Associated Variable		
Active level	Active treatment	
Inactive level		Inactive treatment

The two treatments thus differ on two variables. They are made to differ in only one variable, the independent variable, as follows:

EXPERIMENT WITH CONTROL FOR ASSOCIATIVE CONFOUNDING

	Independent Variable	
	Active level	Inactive level
Associated Variable		
Constant active level	Experimental condition	Control condition

In other words, the experimenter must *actively* put the associated variable at the same level. After seeing how difficult it was to pinpoint the effective independent variable, we should by now be convinced that the pristine experiment is impossible, as Hardin (1972) advised us. It takes a lot of ingenuity to devise an experiment that represents what would happen if one were in fact able to merely do one thing.

SOURCES OF EXPERIMENTAL HYPOTHESES

For applied studies, experimental hypotheses quite often arise from immediate needs or at least possibilities for improvement. They spring directly from the real world. That was certainly the case for the experiments on noise and weaving, on methods for memorizing piano pieces, and on night visual approaches, just to name a few. It is more common for scientific studies, those which seek to improve understanding of behavior, to be based on previous experiments. These could be follow-ups of applied experiments, in such ways as were suggested earlier for pinpointing the reason that unaided search is superior to that with binoculars. More often, they arise from previous scientific research.

Even when one gets an original bright idea, he or she should be somewhat wary of launching immediately into an experiment; somebody else probably got there first. There is not much purpose in testing a possible factor in behavior that has been tested before. Hence, it is necessary to review the *literature*, as the published work on a problem is called. The literature can be roughly divided into two classes: primary sources and secondary sources. Because of the vast bulk of the primary sources in psychology (some 500 journals, at least, some of which have been published for longer than 50 years), it is better to start with the secondary sources. What one should look for is the latest scholarly work on the general topic. Such books are often used as texts for advanced undergraduate courses or for graduate courses. Thus there are books that integrate past research on conflict resolution, on interpersonal perception, on cognitive processes, on biochemical factors in behavior, etc. You will find what the current problems are and obtain many references to the original literature as well. Use also may be made of the *Annual Review of Psychology*, in which 15 or 20 topics or areas are reviewed over the previous year or two. Similar information, but on even more topics, is found in the *Psychological Bulletin*, which is published monthly by the American Psychological Association. Articles in the *Psychological Review*, published every other month by the American Psychological Association, tend to be somewhat more theoretical, with somewhat fewer references, many of the articles being more an expression of original

ideas than reviews of the work of others. Still, to do good experimentation on a problem, you must understand the ideas that are of concern.
Perhaps use of all these sources listed thus far will guide you to problems
that are within your present grasp and let you know what you must learn
to work effectively on other problems.

Once you get into the original studies referred to, you will find that
each of these articles in turn refers to several more articles. You will
know when you have an area covered because after a while you will stop
discovering references to articles that you have not already known about.
Perhaps in order to do this you will have to restrict yourself to reading in
a narrower area than you first were attempting to cover. If you have
access to good library facilities, there will be no difficulty in finding all of
the articles that are relevant. Even without such access, you can usually
obtain a separate reprint of the article by addressing a nice postcard to the
writer. One admonition is, never take some other person's word for what
an article states; you must read it yourself.

OPERATIONAL VALIDITY

The research hypothesis that we wish to test will usually be in rather
general and abstract terms as compared with the concrete operations of an
experiment. The concrete experiment is designed to *represent* the terms
in this hypothesis. In the introduction to this chapter, a question was
raised of whether a loud, raucous buzzer was a good representative of the
more abstract term *punishment*. It would seem like the least typical kind
of punishment. Ordinarily, one thinks of punishment in terms of pain, or
of being deprived of possessions or freedom, or of being subjected to
social disapproval. It is exactly on this question of translation from
abstract or theoretical terms to actual experimental procedures that even
the best of psychologists will sometimes demonstrate questionable logic.
If the translation is shaky, it is said that the experiment is lacking in
operational validity.

An Experiment in Social Psychology

One questionable translation that comes to mind is in the classic
study by Lewin, Lippitt, and White (1939), which was purported to
compare autocratic, democratic, and laissez-faire (i.e., disorganized)
social atmospheres on the behavior of groups of 10-year-old boys. The

problem is in the translation of these concepts to concrete operations. The autocratic atmosphere was represented by an adult leader who made all decisions on activities and made many personal criticisms. The democratic atmosphere was provided by an adult leader who encouraged discussion on activities and was supportive in his approach to the children. For a laissez-faire atmosphere, the adult leader provided no leadership of any kind nor bothered with interpersonal relations. The outcome was that of happier, less bickering, less scapegoating, more productive children under the "democratic" regime than under the others. To this writer, the atmospheres are better described as a despotism, a benign monarchy, and a democracy that perhaps had not had time to work things out (but with an adult "snooper"). Another question on the translation is whether this temporary, part-time life, with a very different atmosphere at home, has anything to tell us at all about individuals who live their lives in one or another atmosphere.

An Experiment on Learning

One of the most eminent psychologists of all time, E. L. Thorndike, tested the hypothesis that frequency of practice can, of itself (i.e., with no knowledge of results), lead to learning (1931, 1932). As his own subject, Thorndike tried to draw lines of a specified length (e.g., 4 inches), over hundreds of trials, while keeping his eyes closed. Unfortunately, he used two contradictory criteria for deciding whether learning had occurred. One was that of performance becoming more accurate; the other was that of fixating more strongly the more heavily practiced early responses. When performance did not become more accurate (see the 1931 study), that was used as evidence against the experimental hypothesis. When performance did improve (1932 study), with the inevitable accompaniment that repetition of early responses became less frequent, that finding too was used as evidence against the hypothesis!

The Completely Appropriate Experiment
Revisited

Were these the *right* experiments to test the experimental hypothesis? Were they externally valid? External validity has previously been gauged by how well an actual experiment represents one of the perfect experiments, the completely appropriate experiment. Thus, it was asked of experiments that "improve" the real world (Chapter 3) that the independent variable, dependent variable, and levels of other variables represent the real situation of concern in the experimental hypothesis. Also, it was asked that when an experiment was conducted on a sample (Chapter 4) the population of interest be well represented. Here the same question is raised, although in a more subtle form.

The issue in regard to the Lewin, Lippitt, and White study (1939) of social atmosphere was whether the three experimental treatments were good representatives of the underlying social atmospheres of the three political systems or philosophies. The objection was raised that any group working with a self-appointed leader could not be considered democratic. However, to insist on the requirement that the leadership must arise entirely spontaneously from the group would mean that the experiment could not be done at all. It was suggested that the laissez-faire atmosphere could lead to a democratic atmosphere if it were given enough time. However, the group might well eventually have an autocratic atmosphere, as in *Lord of the Flies* (Golding, 1954). It is possible to reach the conclusion that although the study fell rather short of being completely appropriate it was about as close to that goal as possible. Thus, external validity perhaps was as high as was practically attainable.

The question of representativeness in that experiment concerned the independent variable; in the experiment on learning, by Thorndike, it concerned the dependent variable. Thorndike was a gifted investigator, so there must have been a good reason why he had difficulty in settling on a dependent variable. He reasoned that without knowledge of results all that mere practice could possibly do would be to strengthen responses that were made at the outset. They would be the ones that were practiced most. Thus, when accuracy of performance did improve, he decided that this change could not represent learning, because he had no way to account for it. However, there are, in fact, other ways learning could take place without *external* knowledge of results. For example, as his subject, he may have learned to attend to his movements more carefully as the experiment went along and, by doing so, to make his responses correspond more closely to his "internal standard" of 4 inches. Theoretical preconceptions should not determine the measurement of behavior used for the dependent variable. The experimental hypothesis was clear enough: Learning means more accurate behavior. A measure of accuracy was the most appropriate dependent variable.

What *operational validity* means is that the concrete experimental operations (manipulation of treatments and obtaining the dependent variable) represent the independent and dependent variables in the impossible, completely appropriate experiment. In this case, the completely appropriate experiment is impossible, because it exists in the world of theory, not in that of concrete operations.

EXPERIMENTAL SUBJECTS

Because of the desire for generality in studies to isolate an independent variable, often there is no specific population referred to in the experimental hypothesis. In other cases, which subjects to experiment on—indeed, which animal—can be inferred from the way in which the hypothesis is stated.

Choice of Animal

Hypotheses can range from the very general to the quite specific in regard to the animal whose behavior is being considered. Sometimes experiments are done for the purpose of finding whether previous findings extend to some different species.

Animals in General   A pigeon will peck increasingly often at a lighted panel if with every tenth peck it receives a grain of corn (just as in the work-ethic experiment). Without the pecking, there is no eating of corn. To put it differently, the eating of corn is contingent on the pecking. The pecking provides the contingency for the eating of corn.

Let us now consider a principle of reinforcement: When a less probable behavior provides the contingency for a more probable behavior, the less probable behavior will gain in probability (Premack, Schaeffer, and Hundt, 1964). Thus, if an animal is deprived of activity but is given plenty of liquid, running about is very probable and drinking is less probable. If the opportunity to run is contingent on drinking, the (non-thirsty) animal will increase its drinking activity. This is meant to be a very general law, applicable to mice and men. We don't know how humble an organism would have to be (protozoa?) not to be implicated. Presumably the hypothesis could then be tested validly on many organisms. If this is so, experimental animals can be selected for convenience. Laboratory animals typically are selected that are small, docile, and survive well in laboratories. Further, they may be bred to enhance these qualities. A fringe benefit of the resulting uniformity is being able to compare experiments from different laboratories. However, there are some unfortunate aspects to this selection of animals. It has been pointed out by Lockard (1968) that the laboratory rat is not a representative animal. It is one that has evolved in companionship with humans and has degenerated from the original stock in many ways. Its very docility results from a stunted system of endocrine glands. This line of argument is supported by the observations by Breland and Breland (1961) on non-laboratory species who do not perform more and more when reinforced more and more. For example, pigs will tend to soon stop bar pressing for food and instead begin to root into the floor, a behavior that is connected with their normal mode of eating.

Particular Animals   Some experiments are conducted to test hypotheses limited to a given species of animal, to males or females, or to young or old. If an experiment is performed to test a factor of homing behavior in pigeons, it is tested on that species. Another reason for using a particular organism is seen in studying the response of cells in the brain when the eye is stimulated by light of different wavelengths. Certain methods are possible only if the study can be done on all-cone retinas, which are found in just a few species, such as ground squirrels (Jacobs and Anderson, 1972). Again, a factor in learning to read can usually be studied only in young children.

People in General   Much of psychology of all kinds, including experimental psychology, is really concerned only with the human species. Thus, there are experiments about people in social groups, about their ability to identify the pitch of a sound, and about their ability to attend to auditory messages to the two ears at the same time. We will not go into the question of whether it is a better strategy to devise a psychology that covers only people or one that covers many species. However, it should be noted that very many experiments on hypotheses directed toward people have been performed on the counterpart of the captive rat, namely the captive student. In some studies, this may lead to overgeneralization of the findings: a special verification of a hypothesis that is taken to be a general verification. It is quite possible that many students at college, because they are living away from home for the first time and because they are at an age when personal alliances are especially important, will succumb to group pressure in an experiment, whereas other persons would not. On the other hand, if the hypothesis is whether or not a certain performance is possible—such as attending to two conversations at once—for *anyone*, then college students may be an excellent choice. They are alert!

Selection and Assignment of Subjects

Once we know the population from which subjects are to be selected, we still must make the specific selections or assignments. In most instances, an available subpopulation will be used, such as the class in elementary psychology. The experimental design used will reflect the investigator's judgment of how best to attain *internal* validity, e.g., to avoid sequence effects or nonequivalent groups. Only rarely will an experimenter who seeks to isolate an independent variable be concerned with selection of subjects to represent a population, which was the main concern of the previous chapter. The reasoning appears to be that it is enough to isolate an independent variable using any subjects; the question of generality is best handled by repeated experiments with different kinds of subjects.

Typical Between-Groups Strategies   *Advance decisions.* If all the subjects to be used in an experiment can be selected from a larger group, the method of random selection can be used for setting up the experimental groups. If all the available subjects are to be tested, the method of random assignment is appropriate. Both designs were described in Chapter 4. Before the experiment starts, it is known which individuals will be in each treatment group.

Serial decisions. Very often the method used for setting up groups is to have subjects choose their own time for appearance among the times available, e.g., through a sign-up sheet. Such a procedure may be employed either when the whole group is used or when only some of the individuals are used. Suppose there are two treatments. Subjects may be assigned alternately to the two treatments as they appear. Care has to be taken, of course, that this does not result in giving Treatment A more or less regularly at a different time of day or day in the week than Treatment B. If there are enough subjects, a random order is more appropriate. A pretest may be used, so that each serial assignment is made to reduce the difference between groups. For example, if, in a reaction time experiment, the mean pretest value for the subjects thus far assigned to Group A was 140 msec while that for present Group B subjects was 150 msec, a new subject with a pretest score of 160 msec would be assigned to Treatment A. It is always a mistake to run all of Treatment A and then run all of Treatment B. The subjects who choose to come around for the experiment earlier or later may well be different in ways that could affect their performance.

Use of Highly Similar Individuals   Comparable groups are sometimes set up in animal experiments by assigning animals from the same litter to the different treatments. Thus, when a number of litters are divided, the different groups are well matched. This is an especially powerful method if a litter consists of animals with the same inheritance, as is the case for opossums. In some species of amphibians, this situation may be produced by the new technique of *cloning*; identical cell nuclei are injected into ova from which the nucleus has been removed. A close approximation to genetic identity exists in highly inbred strains of mice. In research on human beings, perhaps identical twins should be be used more widely in experiments of many types. In this use, it would be all to the good if they were raised together and treated alike. Of course, in studying the relative contributions of heredity and environment, similar upbringing is a confounding factor.

Single-Subject Experiments

A within-subject design may be used in experiments to isolate an independent variable. In fact, this was the case for the experiment on the work ethic. Nothing need be added to the comments on this design in Chapter 2. However, the place of single-subject experiments should be noted.

We started this book with examples of experiments on a single subject and with a very immediate practical goal. This method is also used in more ambitious experiments (Dukes, 1965; Gottman, 1973; Holtzman, 1963). The experimental study of memory, in fact, was started in this way by Ebbinghaus ([1885] 1964) who introduced the use of the nonsense syllable. For most of the factors he studied, transfer effects from the preceding experiment (or two) were probably not too extreme because of the dozens of lists he memorized. One list, more or less, would be of little importance. If we test a hypothesis regarding the existence of an ability, it is sufficient to demonstrate it in one individual. If one person can simultaneously listen to and comprehend two messages at once, this is within human capability. Likewise, if one person shows mental telepathy, knowing what someone else is thinking about, without use of the senses, this capability is demonstrated as being within the range of human ability. For most problems, extreme care must be exercised to achieve good within-subject counterbalancing. In studies of operant conditioning, such as that of Premack, Schaeffer, and Hundt (1964), which was mentioned earlier, a *baseline rate* of response is determined before reinforcement is introduced. It also may be found after the reinforced responses have been extinguished by withdrawal of reinforcement. It is against this baseline that effectiveness of a schedule of reinforcement may be evaluated.

A LOOK BACK AND A LOOK AHEAD

The trend in the preceding chapters of this book has been away from the concrete and the particular and toward the abstract and general. We started by asking which of two treatments works better for one person. The only appropriate generalization was for the future of this person using one of the two treatments. When it was necessary to "improve" the real world to attain internal validity, the experimenter was forced to make decisions on which aspects of the real world it was important to duplicate, i.e., to *abstract* from the real world. Greater generality was achieved in regard to individuals by substituting the sample of a population for the real particular person. Less could be said about what works best for one individual, but more could be said about everyone in the population.

In the present and succeeding chapters, the trend toward abstraction and generalization is continued. In this chapter, abstraction and generalization were closely interconnected. We found ourselves no longer asking about which treatment or level works better than others. Instead, we wanted to answer the question of what it is about treatments that makes a difference in behavior. Let us consider the two parts of this last statement. "What it is about treatments" meant isolating from a complex variable the effective independent variable, which is to say, abstracting the essential aspect. "Makes a difference in behavior" implies less concern than previously over "better than others." This comes about because in scientific investigations "goodness" is usually irrelevant. In a study of learning, an effective variable may be isolated that under one condition brings about rapid acquisition of a foreign language and under another condition rapid acquisition of a neurotic style of behavior. This is analogous to finding the same life processes in cells of the body and in bacteria that cause disease. Thus, the effect of the isolated (abstracted) independent variable may be broadly *generalized*.

Our main interest in the present chapter was on the controls required for isolating an independent variable. We will next turn to the question of deciding whether the observed effect of the independent variable meets certain standards of reliability. Following that, there will be an examination of experiments that specify more fully the way in which independent variables affect behavior. First, there are experiments employing several levels of the independent variable rather than simply pairs of treatments. We shall then see that a clearer account of the action of an independent variable may be obtained through more complex experiments, those with more than one independent variable. It can be seen that a series of steps will have been taken toward the scientific goal of understanding how behavior is determined.

The final topic will be on correlational methods. Thus far the only reference to them was in Chapter 1, in which the correlational approach was compared most unfavorably with experimental manipulation of an independent variable. Still, the correlational method is a way of studying behavior that has made valuable contributions, both practical and scientific. We shall examine the nature of such contributions and consider the possibilities of control.

SUMMARY

The scientific goal of understanding how behavior is determined is furthered by experiments that seek to isolate the effective independent variable. The experimental hypothesis tested concerns a unitary independent variable as compared with the "package deals," which characterize many experiments that are done for immediate application of the results. In these new investigations, the experimental hypothesis is a more demanding one than those considered before, since the independent variable described is one that is not "contaminated" with other variables. An experiment that attempts to isolate a unitary independent variable must be evaluated in respect to internal validity against a very strict standard, the pristine experiment. This experiment is, first of all, an ideal experiment in which the same subject is given different treatments at the same time. It adds the further demands that the independent variable be unitary, not a composite, and that other variables that might affect behavior be held very stable.

It was shown first, through the experiment on Indian girls and the work ethic, that very special circumstances must be set up to obtain treatments that meet the requirements of an isolated independent variable. If the experiment were attempted in a real work environment, for example, it would be impossible to restrict treatments to differences on only the one variable. In some experiments, great skill is required to purify the independent variable. In the example, the study of directional smelling, the investigator contended that the influence of the independent variable, time difference, could be found only by extraordinary efforts to make treatments exact and to eliminate the presence of active levels of other variables, e.g., sounds.

Internal validity is further threatened by some new forms of systematic confounding. They appear to illustrate the sad statement that you cannot do merely one thing. In all cases, they come about because the *active* treatment brings with it the *active* level of another variable. Thus, active level of the independent variable is associated with active level of another variable; inactive level of the independent variable is associated with inactive level of the other variable. In sum, the independent variable is confounded with another variable and the effects on behavior cannot be disentangled. In every case, the cure for this associative confounding was to have the inactive level of the independent variable include the active level of the previously confounding variable. This is called a *control condition*; if a between-groups design is used, there is thus a control group. The active level is now termed the *experimental condition*, and a group thus treated is called the *experimental group.*

The first kind of associative confounding discussed was that of confounding with an *artifactual* variable. The examples were from physiological psychology. If animals who had a part of the brain removed are compared with animals having no operation, the independent variable (removal or not of the part of the brain) is confounded with the artifactual variable (damage to other tissue). Thus there must be an operated control group for comparison with the experimental operated group. This control group is subjected to the same operation except for removal of the part of the brain. Similarly, in an experiment in which introduction of a drug is the experimental treatment, there must be a *placebo* control condition, in which a neutral substance is injected.

Experimentation to test the hypothesis that infants cry as a result of departure of their attachment figure (their mother) was used to illustrate the other kinds of associative confounding. First, there was confounding with a *wider variable*, someone leaving. Second, there was confounding with the *secondary variable*—being with someone or being alone. Third, there was confounding with another secondary variable—whom the child was left with. These are all illustrations of associative confounding, which is called *natural* confounding, in distinction to the *artifactual* confounding described previously. The term *natural* merely means that it is in "the nature of things" for levels of the other variable to be associated with levels of the independent variable, not that it is something that comes about only because of the experimental manipulation. The cure for natural confounding was the same as that for artifactual confounding: obtaining a control condition that includes an active level of the confounding variable. When a diagram is drawn of the levels on which the treatments differ it will be seen that the control condition makes possible a straight-across comparison only on the independent variable.

After a brief discussion of how one goes about looking up previous experiments in order to set up an experiment, the problem of *operational* validity was considered. The question here is that of *external* validity, whether the *right* experiment has been done to test the experimental hypothesis. The variables in experiments to isolate an independent variable are typically rather abstract as compared with the concrete operations of an experiment. Since operations—actual things done—must be concrete, the completely appropriate experiment would be impossible. What must be done is to make the concrete operations as good representatives of the terms in the experimental hypothesis as possible. An experimen on the effect on behavior of social atmosphere was cited, in which there was a question of how well the treatments making up the independent variable corresponded to the terms they were supposed to represent. Then an experiment on learning was questioned, in respect to the representativeness of the dependent variable by the measures of behavior.

The topic of experimental subjects was treated by successively narrowing down the decisions to be made. Some experiments are meant to apply to animals in general, so the choice of species depends mostly on convenience. The question does arise of the generality of the findings, especially when the subjects are laboratory rats. In some cases, a particular species will in fact be chosen because of its particular characteristics. For a large number of experiments, the real interest is the behavior of people. Use of college students is quite prevalent, but the question of generality of results must be examined in the context of the particular experiment.

There still must be some selection or assignment of potential subjects. Typically, the investigator in an experiment to isolate an independent variable does not try to obtain a sample of subjects that represents a population. Instead, he or she will concentrate on questions of internal validity and will deal with the problem of generalization by testing different kinds of subjects in separate experiments. Between-groups designs use one of two strategies. One is to use advance decisions, which was discussed in experiments on representative samples. The other is to use a serial strategy. Alternation of treatments to subjects in order of appearance is advisable when the number of subjects in each group is relatively small. With a larger number, random serial assignment will avoid the possibility of systematic effects. The existence of highly similar individuals may be used to advantage in making groups comparable. Litter mates of some species may be assigned to the different treatments. For human beings, this would encourage experiments using pairs of identical twins, one assigned to each treatment.

Within-subject designs are used about as often as between-groups designs in the present kind of experiment. It is of interest to note that this may even take the form of use of a single subject. In some circumstances, there is effective control of time and sequence effects.

Finally, there was a brief review of the past trends of this book and a preview of subsequent chapters. It was seen that we started with very concrete studies that allowed a minimum of generalization. The movement has been toward the use of more abstract variables and expanded generalization. In subsequent chapters, experiments are described that lead to a more complete, less restricted, account of the factors influencing behavior.

QUESTIONS

1. Compare the experiment on sea rescue search of Chapter 1 with the experiment on the work ethic in regard to the character of the independent variable—unitary or composite.

2. Compare the experiment on preference for brands of tomato juice of Chapter 1 with the experiment on directional smelling in regard to the purity of the experimental treatments.

3. What do we mean by a control group in an experiment that requires removing part of the brain?

4. Distinguish between the procedural confounding that comes about because an experiment cannot be ideal and the associative confounding that comes about because an experiment cannot be pristine.

5. Give an example of natural associative confounding and show how it might be removed. Make tables to describe the situation.

6. How do you go about finding an hypothesis that may be tested experimentally?

7. In experiments to isolate an independent variable, what use is made of the idea of the completely appropriate experiment?

8. How does the experimenter decide on what kind of animal to use in an experiment?

9. What is meant by the serial strategy of assigning subjects to groups?

10. Are experiments to isolate an independent variable usually performed on representative samples? Is this a good idea?

11. Outline the past and future progression of topics in this book.

SIGNIFICANT RESULTS

Let us look at two sets of results obtained by Fleener and Cairns (1970) in the experiment described in the preceding chapter, comparing the crying by babies when the mother leaves the room and when the assistant leaves. For each child there were 24 periods of 5 seconds on which a child was rated when the mother left and 24 when the assistant left. Fifteen children from 12 to 14 months of age cried during 11.67 of these 5-second periods, on the average, when the mother left. When the assistant left, this group cried on 8.27 of the intervals. On the basis of this 3.40 average difference, Fleener and Cairns concluded that the children of this age group cry more often when it is the mother who has left.

In the just younger age group, from 9 to 11 months of age, the means for the 13 children tested were 9.08 when the mother left and 8.15 when the assistant left. Fleener and Cairns (1970) concluded that this small difference, only .93, was not significant. Their conclusion certainly seems reasonable; the difference was very slight for the younger group. But was the difference between 11.67 and 8.27, for the older group, large enough to justify their conclusion of a significant difference? How did they know how large the difference had to be between the two treatments before it could be considered significant?

Their logic was straightforward. They knew that it was possible that the difference for the older group could have occurred by chance. As we have noted, there are many reasons why behavior or its assessment can vary unsystematically from time to time for the same individual and also, on the average, between individuals. In any limited experiment, the chance difference might have made for more crying when the mother left. Yes, a chance difference was possible—but not very *probable*. The investigators were able to *infer* that a difference of this size or more could occur by chance in no more than 1 experiment out of 20. They liked the odds and rejected the idea that theirs was the 1 unlucky experiment in 20.

For the younger group, on the other hand, the difference between 9.08 and 8.15 could have occurred by chance more often than 1 in 20. They thus did not regard the results as showing anything other than a chance difference.

We shall see in this chapter that Fleener and Cairns (1970) were testing the *null* hypothesis—that the experimental treatments make no difference. The word *null* here means "equal to nothing." In the case of the older children, they *rejected* the null hypothesis; in the case of the younger children, they did not. This actually is called *significance testing* or *testing for statistical significance*. When the null hypothesis is rejected, it is said that the difference is *statistically significant*; when the null hypothesis is not rejected, the difference is called (statistically) *not significant*. We shall see that the statistical decision to reject or not always entails two opposite risks. We shall examine the way in which such statistical decisions lead to conclusions on the experimental hypothesis. In so doing, we will find ourselves again involved with internal validity and, more specifically, with reliability. Finally, an attempt will be made to put significance testing in perspective. It will be seen as an aid in drawing a valid conclusion on the experimental hypothesis, but this is far from the whole story. The subject matter of this chapter, *significant results*, thus goes beyond the technical question of statistical significance.

We have contrived to develop this chapter in a somewhat different way than is usual for statistical inference, without equations or calculations. These are to be found, as in previous chapters, in the statistical supplement. Thus, you will not be able to conduct any test of statistical significance unless you go on to the statistical supplement. However, those ideas that are important to experimenters are considered in some detail. If you understand them, it will help you when you read experimental articles, because you will be able to see how conclusions have been drawn. You will know what statistical decisions may be made on the null hypothesis and how they relate to experimental conclusions. Perhaps you will find yourself disagreeing with an investigator either on the statistical decision rule used or on the conclusion drawn from application of the rule.

The main areas on which you will be asked questions at the end of the chapter are:

1. How the null hypothesis is tested.
2. The risks in making a statistical decision.
3. How testing of the null hypothesis is related to internal validity.
4. How this aspect of validity fits into the larger picture of experimental validity.

THE NULL HYPOTHESIS

It seems like a strange approach to test the null hypothesis that the amount of crying does *not* differ when the mother or assistant leaves the room, when that is just the opposite of what the experimenter is proposing. The experimental hypothesis holds that there *is* more crying when the mother leaves. There are two reasons for this "backing in" approach rather than one to prove that the experimental hypothesis is true. The first is that any real experiment—one that is neither ideal nor infinite—cannot be absolutely dependable. We will never be able to say that we have proved, for all time, that our treatments make a difference. We cannot "prove" an experimental hypothesis. The best we can do is to show that *alternative* explanations are unreasonable. This brings us to the second reason for testing the null hypothesis. It is a specific hypothesis and one whose rejection will make sense. Because it is specific (a zero difference between treatments), unlike the experimental hypothesis (any greater amount for one treatment), it is open to standard statistical testing. It makes sense because if we are fairly sure that the treatments do not make "no difference" we are fairly sure that they do make *some* difference.

A Third Possible Conclusion

It might have made you unhappy in the discussion in Chapter 2 to read that in any experiment there is one of two rival conclusions that could be drawn: (1) the experimental hypothesis is supported that the dependent variable has a higher value for Treatment A than for Treatment B, or (2) the counterhypothesis of a higher value for Treatment B is supported. (It is even more likely to have made your instructor unhappy.) What about the conclusion that neither of the rival hypotheses is supported? The point is that we do not have to worry about this third possible conclusion in the use made of the results in simple experiments such as those at the beginning of this book. A weaver would either wear ear defenders or not. (She could not compromise by just wearing one of them if the difference in output was only slightly higher in favor of ear defenders.) If there is no issue of cost or convenience, why not accept any difference, no matter how small? It is always at least slightly more probable that the treatment with the higher value in an experiment will be the one with the higher value in the long run. In other words, in such simple straightforward decisions, the rule to go by is to accept whatever evidence is available. There is no place *in the decision process* for a third conclusion.

However, in the kind of experiment described in the preceding chapter, in which a wrong conclusion will be harmful in developing scientific knowledge in psychology, it is necessary to consider a third possible conclusion, that the independent variable was ineffective. Thus, there were three possible conclusions to be drawn from the results in the Fleener and Cairns experiment, all concerned with what would be found in an ideal or infinite experiment.

1. The hypothesis is supported that children of the given age group would be found to cry more when the mother leaves.
2. The hypothesis is supported that they would be found to cry more when the assistant leaves.
3. Neither hypothesis is supported.

These investigators realized that just by chance in any real, limited experiment the difference could be positive (more crying when mother leaves) or negative (more crying when the assistant leaves). Thus, only a large-enough difference of crying with mother leaving over crying with assistant leaving would be taken as supporting the experimental hypothesis that this would be found in an ideal or infinite experiment. A smaller difference would have too high a probability of being accidental.

We may represent the relation between the difference obtained and the conclusion to be drawn by the following diagram.

Conclusion—Results Support:

Counterhypothesis	Neither hypothesis	Experimental hypothesis
More crying with assistant leaving		More crying with mother leaving

```
  .      .      .      .      .      .      .      .      .      .      .
 -5    -4    -3    -2    -1     0    +1    +2    +3    +4    +5
```

Amount of crying with mother leaving minus amount of crying with assistant leaving

It is seen by the short vertical mark over the value $+3.40$ on the right that this amount of difference (for the older group) was sufficient to support the experimental hypothesis of more crying with mother leaving. On the other hand, the short vertical mark over the value $+.93$ (the difference for the younger group) shows that this difference is insufficient to be taken as support of the experimental hypothesis. A difference of almost 3, plus or minus, is required for support of either the experimental hypothesis or the counterhypothesis.

Rejecting or Not Rejecting the Null Hypothesis

A Statistical Decision Rule   The three possible conclusions to be drawn from the results of an experiment are made on the basis of a *statistical decision rule*. Here it was that the null hypothesis can be rejected only if the probability is less than .05 (i.e., less than 1 in 20) of obtaining as large a difference as that found if the null hypothesis were true.

The Basis of Statistical Inference   If Fleener and Cairns had replicated their experiment (done it over and over) with other groups of children drawn from the same age group, they would not have obtained the exact difference of 3.40 on each experiment between the means for mother leaving and for assistant leaving. Because of unsystematic variation, it would sometimes be higher, sometimes lower. If for an infinite number of replications the overall mean difference was exactly 0, this would mean that the null hypothesis is true. Still, for any particular one of the experiments, some other value than zero would be expected.

The mother-minus-assistant difference will, then, vary from experiment to experiment. The amount of variation among these differences depends on the reliability of each experiment. As we saw in Chapter 2, there is better reliability—and less variation to be expected from experiment to experiment—the larger the number of observations and the smaller the amount of unsystematic variation. Hence there would be less variation in the mother-minus-assistant difference if each experiment used many subjects and had a small standard deviation.

From the number of subjects tested and the standard deviation among them, it is possible to *infer* the amount of difference that would be exceeded with a probability of .05 if the null hypothesis were true. This is the process called *statistical inference*. That value for the Fleener and Cairns experiment was not quite ± 3. (This determination came from a statistical test called the *t-test*; you may refer to the statistical supplement to this chapter. It is one of the many tests of statistical significance used by experimenters.)

We see in the diagram on page 195 how a *statistical decision rule* was applied to draw one of the three possible conclusions from the Fleener and Cairns experiment.

The difference of +3.40 for the older children is seen to fall into one of the two *rejection regions*. If the null hypothesis were true, there would be a total proportion of only .05 experiments showing a difference in one or the other rejection region. The probability (*p*) for a given experiment would be .025 for each rejection region and .95 for the "do not reject" region. Using a .05 decision rule, we say that the difference is *significant*, since we are able to reject the null hypothesis. The smaller difference of +.93 is seen not to fall in a rejection region. The statistical decision is

Conclusion—Results Support:

Counterhypothesis More crying with assistant leaving	Neither hypothesis	Experimental hypothesis More crying with mother leaving

```
 •      •      •      •      •      •   |  •      •      •      •      •      •
−5    −4    −3    −2    −1     0    +1    +2    +3   ⸝+4    +5
```

Statistical Decision

Reject null hypothesis ($p = .025$)	Do not reject null hypothesis ($p = .95$	Reject null hypothesis ($p = .025$)

```
 •      •      •      •      •      •   |  •      •      •   |  •      •      •
−5    −4    −3    −2    −1     0    +1    +2    +3    +4    +5
```

Amount of crying with mother leaving minus amount of crying with
assistant leaving

thus not to reject the null hypothesis. This result could have been
obtained with a higher probability than .05 if the null hypothesis were
true. For the older children, we may conclude that there was more crying
with the mother leaving. We cannot draw this conclusion (or the opposite
one of more crying with the assistant leaving) for the younger group.

Factors Affecting the Amount of Difference Required

The preceding diagram has shown the amount of difference between
means required for rejecting the null hypothesis given just one particular
set of circumstances. With more reliable data, a smaller difference would
suffice for rejecting the null hypothesis. However, if a stricter criterion is
used for rejection of the null hypothesis, a larger amount of difference
between means will be required. These two factors are illustrated in
Figure 6.1.

Effect of Reliability   If either more children had been tested or if the
standard deviation within each group had been smaller, reliability would
be improved. What this means is that in the set of infinitely replicated
experiments the means would not vary as much. If such is the case, the
difference between means would not vary as much. There would be a
"tighter pattern" around the true overall mean. Thus, if the null hypoth-

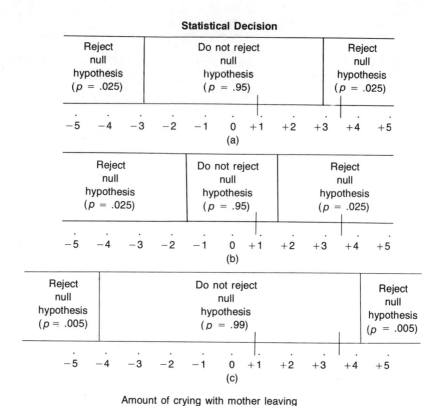

Amount of crying with mother leaving
minus amount of crying with assistant leaving

Figure 6.1 Original statistical decision shown in (a); effect of increasing the reliability (b) and of using a lower alpha (c) on the amount of difference required for rejecting the null hypothesis.

esis were true, the two lines showing where, plus or minus, .025 of the actual means would fall would be closer to zero.

In Figure 6.1, the top diagram (a) is that already shown, in which there were 15 children tested. In the middle diagram (b), the lines are shown for the rejection regions if the number of children had been increased to 60 or if the standard deviation for each of the treatments had been reduced by one-half. In either case, the required difference is cut in half, to just a little under ±1.5. It is seen that the .93 difference for the younger group still is out of the rejection region, again disallowing support for the hypothesis of more crying with the mother leaving. However, if reliability were increased by a great amount, either by using a much larger sample or reducing variability, even a .93 difference—*if found*—would be significant. The point is that with more reliable data, it would be less probable to obtain that size difference if the null hypothesis were true.

Effect of Decision Rule   Up to now, only one decision rule has been used. The null hypothesis is rejected if the probability is less than .05 of obtaining the difference if the null hypothesis were true. The probability used for rejecting the null hypothesis is called the *alpha level*. If a stricter criterion is used, it means rejecting the true null hypothesis some smaller proportion of time. Most often this is set at .01 (1 in 100 experiments) instead of .05 (1 in 20 experiments).

The effect of using an alpha level of .01 in the decision rule, instead of .05, as previously, is seen in the bottom diagram (c) in Figure 6.1. (The original sample of 15 children and original standard deviations still hold.) With the .01 alpha level a difference of more than ± 4 is required to reject the null hypothesis. With this stricter test, we cannot conclude that the experimental hypothesis of more crying with the mother leaving is supported.   ·

The way in which a statistically significant difference is reported in most articles is "$p < .05$" or "$p < .01$." This means probability in respect to the null hypothesis was less than .05 or .01. A nonsignificant difference is reported as "$p > .05$" or "$p > .01$."

FROM DECISIONS TO CONCLUSIONS

People who do not like statistics may feel that nothing much is accomplished by all this fancy testing of statistical significance. They are wrong. Other people who worship statistics may feel that for each statistical decision there is an automatic experimental conclusion. They are wrong, too. We shall attempt to show that a position between these two extremes makes sense. Motto: Use statistics, but make it your servant; do not let it be your master.

Failure To Use a Significance Test

Suppose that Fleener and Cairns had disregarded the null hypothesis. They were willing to take *any* amount of difference of more crying with mother leaving as support of their experimental hypothesis. On this basis, they would have taken the .93 difference for the younger group as being a significant difference. This sounds like risky business. In further examining their article, it may be noted that for the youngest group studied (from three to five months of age) there was a difference of 1.78 in the opposite direction, more crying with the *assistant* leaving. Thus, by always rejecting the null hypothesis they would have found themselves supporting the counterhypothesis, which in this case is most unreasonable.

We can immediately see the effect of never rejecting the null hypothesis. When the null hypothesis happens to be true, experimental conclusions will always be wrong. Either the experimental hypothesis or the counterhypothesis will be considered to be supported. Moreover, when the null hypothesis is false and there is some true difference in one direction or the other, a false conclusion will often be made. Suppose that for the youngest children, over an infinite number of experiments, there really would have been a slight difference in more crying with the mother leaving. In any particular experiment, the difference might well come out the other way. This latter error will be made the more often the poorer the reliability of the data. As a matter of fact, only nine children made up the youngest age group, so the reliability was, in fact, suspect. In sum, we cannot get by in a scientific experiment without significance tests.

Risks and Types of Error

Since we know that actual experiments are neither ideal nor infinite, we know that some of our decisions will turn out to be in error, no matter what decision rule is used. Perhaps Fleener and Cairns should not have rejected the null hypothesis for their oldest group; perhaps the null hypothesis is true. If they had used the .01 alpha level, they would not have been able to reject the null hypothesis. That would be just fine if, in fact, the null hypothesis were true. But what if it were not? With either alpha level, they would take a risk, but of opposite kinds.

Type I Errors The first risk is that of making a *Type I error*: rejecting the null hypothesis when it is true. If an investigator uses the .05 level for the decision rule, it means that he is willing to make this error in up to five percent of his experiments. Since the investigator takes the rejection of the null hypothesis as support for the experimental hypothesis (e.g., more crying with mother leaving), this means that he is being rather optimistic. There is 1 chance in 20 that the claim of an effective variable is false! In any experiment to test an absolutely new hypothesis that runs counter to general thinking, more caution is advisable. It is a very serious thing to upset the applecart of science, and one should be awfully sure of one's facts. In such a case, the more stringent decision rule, use of a .01 alpha level, is more advisable. Science can probably tolerate a 1 percent rate of results that wrongfully claim to support an experimental hypothesis, but not a 5 percent rate of false claims.

Type II Errors   If we insist on the .01 alpha level (or an even more stringent level, such as .001), the risk becomes substantial that our insistence on being almost absolutely sure has led us to fail to reject the null hypothesis when it is false. Naturally enough, this is called the *Type II error*. If the null hypothesis is false, some other hypothesis must be true. The risk of failing to reject the null hypothesis when this other hypothesis is true (e.g., some particular difference in crying, mother leaving minus assistant leaving), can also be stated as a probability, called *beta*.

For a given set of experimental results, reducing the alpha level means increasing the beta probability for any nonzero hypothesis. Use of a very stringent decision rule means that the experimenter is willing to accept considerable risk of failing to reject the null hypothesis when some other hypothesis is true. Thus, with a low alpha level, the experimenter will often wrongfully conclude the experimental results did not support the experimental hypothesis. Unlike the alpha level, no overall probability number can be given for the beta probability; it differs for each particular nonzero hypothesis of the difference between treatments. Thus, if the hypothesis of a large difference between treatments happens to be true —say, +5.0 in amount of crying—the probability is *low* (beta) of failing to reject the null hypothesis, even with use of the stringent .01 alpha level. On the other hand, if the true difference happens to be a small value—say, +1.0—the probability will be much higher of wrongfully failing to reject the null hypothesis. Still, the logic holds up. With the same data, reducing the alpha level increases the beta probability for all statistical hypotheses other than the null hypothesis. A statistical test of experimental results is said to have *power* to the extent that beta values are generally low for nonzero hypotheses. With good power, real differences are identified. Of course, power is automatically increased by using a lax decision rule (e.g., a .10 alpha level), but that increases the risk of a Type I error. There are two better ways of increasing power. One is to have more reliable data. As we saw in Figure 6.1(b), either increasing the number of subjects or reducing unsystematic variation allows rejection of the null hypothesis for a smaller difference between treatments. The other way is to use the most effective experimental designs and statistical tests. This last point is developed in more advanced tests (e.g., Cohen, 1977).

In the previous section, it was stated that the Type I error should be avoided when general beliefs or results of previous experiments were being challenged by rejection of the null hypothesis. On the other hand, if the experimenter finds no significant difference between treatments that are generally believed to be effective, this conclusion should be made on the basis of a *high* alpha level, so as to reduce the risk of a Type II error. Almost any correct previous finding can be "overturned" through erroneous failure to reject the null hypothesis either through the use of unreliable data, use of a stringent decision rule, or (*worse*) through both. Let us now look into the matter of what the experimenter ought to conclude with rejection of the null hypothesis.

Conclusions When the Null Hypothesis Is Not Rejected   Please note—only one of two statistical decisions is made in respect to the null hypothesis: to reject it or not to reject it. *Nowhere* is there a decision to *accept* the null hypothesis. Still it makes good sense for the experimenter sometimes to *conclude* that the independent variable had no effect. In the diagram on page 194, the conclusion supported with failure to reject the null hypothesis is that neither the experimental hypothesis nor the counterhypothesis was supported. For example, for the younger group of children, the small difference in amount of crying favored neither the hypothesis of more crying with mother leaving or the counterhypothesis of more crying with assistant leaving. However, there are different conclusions that may be drawn from nonsupport.

First, an experimenter may conclude that he honestly does not know whether the independent variable made any difference in behavior. This interpretation is especially fitting if reliability had been low because only a few subjects had been tested or if behavior turned out to be more variable than was expected. Thus, Fleener and Cairns might have decided to do a follow-up experiment on the younger group and attempt to reduce unsystematic variability as much as possible.

Second, an experimenter may conclude that reliability was quite satisfactory and that the failure to reject the null hypothesis really did mean that the treatments made no difference. This interpretation would be bolstered if earlier experiments had indicated the independent variable to be ineffective. The statistical decision has again been not to reject the null hypothesis. However, from the circumstances of the experiment, this has led to the conclusion that the independent variable is ineffective.

The Validity of Conclusions

Let us go back to the definition of internal validity given in Chapter 2: the extent to which the conclusion on the experimental hypothesis may be depended on to be the same as would follow from an ideal or infinite experiment. In the preceding chapters, we have seen how internal validity is strengthened by procedures that give reliable data and that avoid confounding. It is clear that conclusions cannot be better than the data. In the present chapter, we have seen how the thoughtful use of statistical decision rules leads to justifiable conclusions on the experimental hypothesis. This, too, is a way of strengthening internal validity, since the conclusion drawn is a most important part of the reported experiment. We may develop this line of thought in a fairly direct way.

The infinite experiment both defines complete internal validity and provides the basis for testing the null hypothesis. Of course, for the latter purpose, it is a particular kind of infinite experiment. It is broken up into separate experiments. Each of these is just like the experiment that has actually been conducted, but uses different subjects drawn from the same population (or different trials, if we are referring to a within-subject design).

In testing the null hypothesis, we are asked to suppose that the conclusion that would be flawlessly drawn from this infinite experiment is that the experimental treatments made no difference. Naturally, the overall average difference between treatments from this assemblage of subexperiments would be zero. However, this would not be the case for each of the subexperiments. The difference between means would only *center* around a value of zero, but some experiments would favor one treatment and some the other. Now we are asked to consider the difference we obtained in our own experiment in relation to the whole *range* of differences that would be provided by this kind of infinite experiment.

Dependable Conclusions if the Null Hypothesis Is True   If the null hypothesis happens to be true—if the overall mean difference between treatments in the infinite experiment is zero—we wish to be able to arrive at that conclusion from our experiment. We do not wish to conclude in favor of an hypothesis of a difference between treatments if there is much chance at all that on the basis of the infinite experiment we would conclude there was no difference. Thus, from reliable data, we will conclude that the experimental hypothesis that the treatments did make a difference is incorrect if there is a probability of .05 or .01 that a difference as large as ours could have come about if the null hypothesis were true. We would most strongly want to *depend on* reaching that conclusion when testing a novel experimental hypothesis, perhaps one that goes against general knowledge or beliefs. Thus the alpha level is set at .01 or lower in such a case. With the .05 level, 5 percent of the conclusions will represent false claims. In an infinite number of experiments, 5 percent would produce the results that led to the rejection of the null hypothesis in the single experiment.

　　　Dependable Conclusions if the Null Hypothesis Is False　　If the null hypothesis happens to be false—if the overall mean difference between treatments is as predicted by the experimental hypothesis—we would like to draw that conclusion from our experiment. This is not so important in testing the novel idea. With further experimentation, its time will come if it is correct. However, if a difference between treatments *would* be expected on the basis of current knowledge, we want to be able to *depend on* our conclusion to be in favor of the experimental hypothesis.

　　　As we have seen, this requires reliable data. In addition it calls for use of a less stringent decision rule, e.g., use of the .05 alpha level. If the null hypothesis happens to be true, we still would like to reach that conclusion. However, we are willing to increase the risk somewhat of incorrectly rejecting the null hypothesis, in order to reduce the risk of concluding that the experimental treatments make no difference if, in the infinite experiment, they really would.

STICKY PROBLEMS THAT REMAIN

An experimenter may have tiptoed safely through the minefield of risks in drawing conclusions in respect to the null hypothesis and still fail to make much of a contribution to knowledge. In this section are presented three "sticky problems" that threaten the internal validity of conclusions based on decision rules, even though data are reliable and significance tests have been interpreted thoughtfully.

Strong and Submerged Effects

　　　One waggish scholar has suggested that a good procedure to use in deciding about the significance of an effect is the interocular traumatic test: "You know what the data mean when the conclusion hits you between the eyes" (J. Berkson, reported in Edwards, Lindman, and Savage, 1963). He must have been talking about an experiment in which there is an expectation of a *strong* effect, a large difference between treatments.

When an independent variable is studied, there is usually one of two kinds of expectation. The first is that the behavior in question depends strongly on the variable and that its presence or absence makes a great difference. Thus, we would expect that a blindfolded person could localize the direction of a sound well only if there were a physical difference in the sounds reaching the two ears. If there was fairly good success without this difference, it would mean that the independent variable was not nearly as important as was thought. The experimental hypothesis would not be supported even though the null hypothesis was rejected.

On the other hand, a small but consistent difference is sometimes all that reasonably can be expected from the different treatments. Consider an experiment on perceptual defense, said to be demonstrated by failure to identify obscene words presented with short exposure. The difference between obscene and neutral words in number identified (whatever the reason) would not be expected to be large. This is because the effect is almost bound to be *submerged* by other factors. There might be neutral words that are seen or reported inaccurately because of shapes of the letters or confusion with other words. Also, some of the "neutral" words may refer to embarrassing experiences in the particular individual's past. Finally, there are ups and downs in alertness during a session, which probably affect identification more than do the variations in "cleanliness." With this expectation, a slight but consistent tendency toward poor identification with the obscene words would suffice to show that treatment made a difference.

The Fleener and Cairns experiment was one in which the effect could be considered submerged. Although there may be a natural tendency toward maternal attachment and thus toward more crying with the mother leaving, the effect might well be obscured by other factors. Some mothers, because of work, may be with their infants a relatively small amount of time with the attachment not developing its potential strength. Other mothers, because of personal habits, may dart in and out of the infant's view countless times each day, tending to accustom it to disappearances. Assistants will resemble mothers more or less in appearance and in manner. Further, it was found by Fleener and Cairns (1970, p. 218) that some of the infants cried most of the time during the experiment: "Possibly the most obvious feature of the crying was its persis-

tence: If the infant began to cry vigorously, he was likely to continue."
All of these factors would make it difficult to demonstrate a difference
between mother leaving and assistant leaving. This experiment is thus
one in which a strong effect could not be expected. The variable of
person leaving may be shown adequately by a statistically different
amount of crying. It would not require a difference that "hits you between the eyes."

No Safety in Numbers

A curious thing happens as the number of subjects in an experiment
is greatly increased, thus increasing the power of the significance test.
Any two treatments, it seems, will then provide a difference that is statistically significant (Bakan, 1967). If significance is not found for 20
subjects, it will be for 200 or 2,000 or 2,000,000. The reason is probably
not mysterious at all. Any two treatments compared involve many factors
other than those which they were intended to exemplify. We have already
pointed out that it is beyond human capacity to control for every conceivable associated bias. Perhaps letters are recognized better than digits only
because there are a few subjects in every thousand who have reacted emotionally to unhappy experiences in mathematics courses. Perhaps words
spoken every 8 seconds may not be heard quite as well as those at some
other interval because, over all, they coincide slightly more often with the
usual swallowing cycles (an act that reduces ability to hear).

Such *trace factors* will make their presence known when mountains
of data are analyzed. The main lesson in this is that we should not allow
ourselves to become overdependent on the significance test in reaching
conclusions about hypothesized factors. It is only one tool. Specifically,
we should be somewhat dubious that the independent variable produced
the obtained difference when it required so many data to bring out the
effect. We should really be impressed with statistically significant differences found more easily, with relatively few subjects or trials.

Does the Conclusion Hold for All Subjects?

It would be possible for Fleener and Cairns to obtain their results for
the older group, which proved to be statistically significant, even if only
nine of the fifteen children showed more crying with the mother leaving.
If this happened, what would we say about the other six?

We expect a real psychological factor to be effective for every subject in an experiment. And you will find, in reading articles in the journals, that such an attribution is almost always made. The unspoken assumption is that when some independent variable is truly effective it works for all individuals covered by the hypothesis. If it truly holds for some, it holds for all. The reason that there was not more crying with the mother leaving for the six negative cases is held to lie in the kinds of submerging factors that were previously mentioned: past experience with the mother, previous crying during the experiment, etc.

This may not always be the case. Let us consider another experiment. Suppose it is found that subjects are able to recognize words better at a later time if they say them out loud at first sight. At least this is what happened for 13 out of 20 subjects. Now, the reason the results did not hold for the other seven subjects might simply lie in unsystematic variations such as previous associations with some of the words used. However, it may be that some people are aided in recognition by immediate recitation, but others are not. Using a within-subject design, extensive testing could be done on each subject to explore the possibility of real individual differences. Even better, perhaps we can find some characteristic that distinguishes between those who are aided and those who are not. It could be that people who visualize extremely well obtain no added advantage from recitation. But we are getting ahead of ourselves on a matter that will be considered in Chapter 8, where we witness a new control or two being born.

OTHER ASPECTS OF VALIDITY

In this chapter, we have talked about conclusions based on statistical decisions. However, we should not lose sight of the fact that there are other important aspects of validity. Too often a conclusion is held to be valid entirely on the basis of considerations of reliability, which have occupied our attention in this chapter. We, of course, know that there is more to validity than that.

External Validity

We will recall, first of all, that an experiment may lack *external* validity for several reasons. The experiment might not represent the *completely appropriate* experiment because of the wrong level of another variable (if Jack Mozart memorized waltzes instead of sonatas when he was comparing methods of practice). In the experiments that improve on the real world (e.g., simulating night visual approaches), we wanted to also make sure that the artificial independent and dependent variables were representative of those situations to which the results would be applied. When a sample of subjects was tested (as in the experiment on methods of providing price information), we were concerned with how well it represented the population of supermarket shoppers. When concrete procedures were devised to represent the theoretical ideas of social atmosphere (autocratic, democratic, or laissez-faire), we did a certain amount of headshaking about the *operational* validity of these procedures. All of our statistical decision making is silent in respect to external validity. Still, experimental conclusions cannot be fully valid unless they have external validity as well as internal validity.

Systematic Confounding

Second, we will recall the emphasis in Chapter 2 on avoiding systematic procedural confounding (such as comes about by sequential effects) and in Chapter 5 on avoidance of associative confounding. We have seen that Fleener and Cairns (1970) concluded that there was more crying with the mother leaving the infant than with the assistant leaving. We have previously seen that Cohen (1977) observed that there was systematic associative confounding. The *staying* person was different when the mother or assistant left. Thus there is a cloud on internal validity no matter how large the difference is between the two treatments. Rejection of the null hypothesis has nothing to say about systematic confounding. Experimental conclusions can have internal validity only to the extent that systematic confounding has been avoided.

When we talk about the validity of conclusions drawn from experimental results on the basis of statistical decisions—used well or poorly, or wrongfully disregarded—we must assume that the foregoing aspects of validity are satisfactory. You should keep this in mind (or maybe write it on the palm of your hand).

SUMMARY

A large difference in the effect of different treatments will lead an experimenter to conclude that the experimental hypothesis has been supported. Some smaller difference will be interpreted as a chance result. The basis of the different conclusions is statistical significance. More specifically, this means that if there were no difference in an ideal or infinite experiment it would be improbable on a particular experiment to obtain the large difference but not so improbable to obtain the smaller difference.

In scientific experiments—as contrasted with those in which there are only two courses of action—there are three possible conclusions that may be drawn from the data. In addition to the conclusions that the experimental hypothesis is supported or that the counterhypothesis is supported, there is also the possible conclusion that neither one is. One of these three conclusions will be drawn on the basis of the statistical decision made on the null hypothesis.

If countless experiments were performed, and the null hypothesis were true, the average difference between treatments would be zero. However, the difference for individual experiments would sometimes be in one direction and sometimes in the other. If it is inferred that a difference as large as the one obtained would seldom occur in the infinite experiment, the null hypothesis is rejected. However, if the probability of occurrence of a difference as large as that obtained is too high, the null hypothesis is not rejected. When the null hypothesis is rejected, it is concluded that support has been given to the experimental hypothesis (or to the counterhypothesis, if the difference was in the unexpected direction). When the null hypothesis is not rejected, neither the experimental nor the counterhypothesis is supported. This latter conclusion may mean one of two things. If the data are unreliable, the conclusion will be that it simply was not established that the independent variable was effective. With reliable data and previous evidence of no effect, the experimenter may be justified in holding that treatments made no difference.

Two factors determine how large a difference between treatments is required to reject the null hypothesis. First, there is reliability. With greater reliability, a smaller difference will suffice for rejection. The second factor is the probability that the experimenter sets for the risk of falsely rejecting the null hypothesis when it is true. This is called the *alpha level* for his decision rule. The error that will be increased as this risk is made larger is called the *Type I error*. Thus, the risk of a Type I error is five times as high with a .05 alpha level as with a .01 alpha level.

However, there is a greater risk of the opposite error with the lower alpha level. It is that of failing to reject the null hypothesis when some other hypothesis is true (and, of course, the null hypothesis is false). This is called the *Type II error*. For any given set of data, this probability—called *beta*—increases as the alpha level is reduced. However, by making the experiment more reliable, reasonable beta values will be found even with stringent alpha levels. A statistical test is said to have *power* to the extent to which the beta probability is low, the extent to which a true difference will be identified.

Use of a strict alpha level (e.g., .01) is recommended when difference between treatments would support an hypothesis that is contrary to general thinking. That is because science should not be forced to deal with too many nonfacts. If 5 out of 20 positive conclusions were wrong, it would be a heavy burden on the scientific enterprise. On the other hand, where there is previous evidence for an effect of an independent variable, it should not be discounted because the difference failed to achieve a .01 level of significance.

The function of significance tests is to further internal validity. This idea was developed by pointing out that internal validity and tests of the null hypothesis may both be described through the infinite experiment. In the infinite experiment consisting of subexperiments (such as that actually performed), the overall mean difference between treatments will be zero if the null hypothesis is true. However, the differences found in the subexperiments will vary around zero. The experimenter is able to infer how they will vary. He considers the difference he obtained relative to this variability. He will not conclude that the treatment made a difference if too many subexperiments give a difference as large or larger. Should the null hypothesis be true, the experimenter wishes to reach that conclusion *dependably*. However, some risk must be taken in order ever to conclude that some other hypothesis is true. The experimenter wishes dependably to reach the conclusion that the experimental hypothesis is supported should there really be the expected kind of difference in the infinite experiment. His balance between the risks of Type I and Type II errors reflects his appraisal about relative importance of the two kinds of dependability.

Three sticky problems remain in drawing conclusions. The first is that a merely significant difference may not be enough when the independent variable would be expected to provide a strong effect. Statistical testing is most applicable to cases of effects submerged by other factors acting unsystematically. The second problem is that using too many subjects will allow certain trace factors to become evident. The third problem is that of the universality of effects. How can conclusions be applied to all individuals of concern when they are not found for all subjects tested? It may not be the case that this is due only to unsystematic variation. Finally, it was pointed out that we cannot accept experimental conclusions solely on the basis of consistent and strong enough differences between treatments. The experiment will still lack external validity if it is inappropriate in any one of a number of ways. Moreover, it will not even be internally valid if there has been faulty control of systematic confounding.

QUESTIONS

1. Why did Fleener and Cairns conclude that older children cry more when the mother leaves than when the assistant leaves, but that there is no difference for younger children?

2. What is the null hypothesis?

3. Why is there a third possible conclusion in the Fleener and Cairns study, but only two in Yoko's experiment on which brand of tomato juice she preferred?

4. What is shown in the diagram that indicates difference between treatment means, statistical decision, and conclusion on the experimental hypothesis?

5. What is the effect of decreased reliability on the amount of difference between means required for rejecting the null hypothesis?

6. What is the effect of alpha level in the decision rule on the amount of difference between means required for rejecting the null hypothesis?

7. Relate alpha level to risks of Type I and Type II error.

8. When is it of special importance to avoid Type I error?

9. Describe three factors that influence beta probability. What does that mean in respect to risk of a Type II error?

10. Under what conditions may an experimenter conclude that the independent variable is ineffective?

11. How can it be said that the thoughtful use of statistical decision rules contributes to internal validity?

12. Can there be too many subjects in an experiment?

13. If an experiment has produced reliable data and a large significant difference is found between treatments, is it assured that the conclusions are satisfactorily valid?

 STATISTICAL SUPPLEMENT

THE t-TEST

In this supplement, a method will be shown for finding the amount of difference between means necessary for rejecting the null hypothesis. In effect, we will be developing the diagrams shown in Figure 6.1.

A Sampling Distribution

Let us again suppose that the reaction-time data presented in the previous statistical supplements are those of a between-groups experiment. We thus have a mean reaction time for each of the 17 subjects given Treatment A (light) and a mean reaction time for each of the 17 subjects given Treatment B (tone). Moreover, we have an overall mean for the subjects given Treatment A (185 msec) and an overall mean for Treatment B (162 msec). Finally, we have a difference between these two means, $M_A - M_B$, of +23 msec.

If two more groups of subjects, selected in the same way, were tested, we would not expect $M_A - M_B$ to equal 23 msec *exactly*. Nor would we expect the difference in a third experiment to be *exactly* 23 msec. Rather, we expect to find the value of $M_A - M_B$ to vary from experiment to experiment because of unsystematic variation.

Suppose the infinite experiment were realized by replicating this experiment, 17 subjects being given each treatment countless times. Let us further suppose that the null hypothesis is true. Then the difference between the population means—which is a *parameter* —would be zero. In other terms, $\overline{M}_A - \overline{M}_B = 0$. However, the value of the statistic $M_A - M_B$ would vary from experiment to experiment.

A distribution of the values for $M_A - M_B$ could be drawn for the successive experiments as described. Let us represent the value found in the experiment that gave the difference of +23 as 1; the value that might be found in a second experiment (say, −4) as 2; the value for a third experiment (say, zero) as 3; etc. Thus for nine experiments, we might find, if $\overline{M} - \overline{M}_B = 0$:

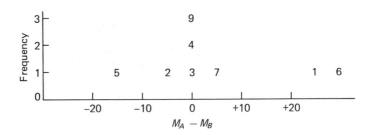

Fortunately it is possible to *infer* what this distribution would look like for the countless experiments. We can actually draw the distribution of values of $M_A - M_B$ that would be expected to occur.

Moreover, we can estimate the standard deviation that this distribution would have. This kind of inferred distribution is called a *sampling distribution*. Since the distribution being described is that of the difference between means, the present distribution is the sampling distribution of the difference between means. (There are also sampling distributions for means, for standard deviations, etc.).

This is the sampling distribution for our reaction-time experiment, assuming the null hypothesis, $\overline{M_A} - \overline{M_B} = 0$, to be true.

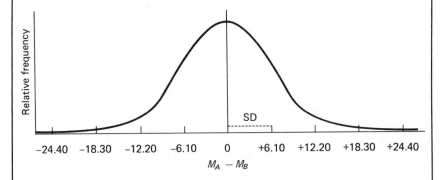

You will note that the estimated standard deviation (SD) is 6.10. Thus an $M_A - M_3$ difference of $+12.20$ for an experiment lies two statistical deviations above the assumed $\overline{M_A} - \overline{M_B} = 0$ value, an $M_A - M_B$ difference of -18.30 lies three standard deviations below the assumed zero value, etc.

The Standard Error

It still has not been explained how the standard deviation of this hypothetical sampling distribution was computed. This is the formula:

$$s_{M_A - M_B} = \sqrt{\frac{s_A^2}{N_A} + \frac{s_B^2}{N_B}} \qquad \text{(Formula 6.1)}$$

The name for $s_{M_A - M_B}$ is the *standard error of the difference between means*. The term *standard error* rather than *standard deviation* tells us that we have inferred the standard deviation rather than going through the (impossible) infinite calculations. You will note that s is used rather than σ. This is because a population parameter ($\overline{\sigma}_{M_A, -M_B}$) is being estimated from sample statistics.

To do the computation, simply insert into the formula the values of s_A^2 and s_B^2, which have already been obtained in previous statistical supplements. Thus:

$$s_{M_A - M_B} = \sqrt{\frac{363}{17} + \frac{269}{17}} = \sqrt{21.35 + 15.82}$$

$$= \sqrt{37.17}$$

$$= 6.10 \text{ msec}$$

You can see that this formula may also be used when N_A and N_B differ, that is when there is a different number of subjects (or trials in a within-subject design) for the two treatments.

Finding the Value of t

The next step is to find how many standard-error units our obtained $M_A - M_B$ lies from the null hypothesis mean of zero. Since our obtained difference was +23, and the standard error of $M_A - M_B$ was 6.10, it is evident that our difference was 3.77 standard error units above zero. The standard error units are called t *units*. Expressing an obtained difference in terms of standard error units is called *finding the value of t for the difference*. This can be expressed by the formula:

$$t = \frac{(M_A - M_B) - 0}{s_{M_A - M_B}} \qquad \text{(Formula 6.2)}$$

Substituting the values from our reaction time experiment, we have

$$t = \frac{(185 - 162) - 0}{6.10} = \frac{23}{6.10}$$

$$= 3.77$$

You may note that the zero in the numerator may be disregarded in the computation. It only serves to remind us that we are testing the null hypothesis, $\overline{M_A} - \overline{M_B} = 0$.

Rejecting or Not Rejecting the Null Hypothesis

We are now ready (at last!) to describe how the diagrams in Figure 6.1 were obtained, showing the amount of difference between means required for rejecting the null hypothesis. Let us redraw the sampling distribution of the difference.

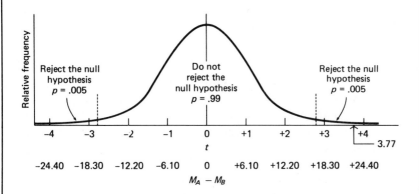

You will find in Statistical Table 2, at the end of this supplement, the value that t must have to reject the null hypothesis. This is given for both the .05 alpha level and the .01 alpha level. These *critical values* depend on the value of N for each treatment or, rather, on the number of *degrees of freedom*, $N - 1$, for each mean. (If you have a given mean, say 179 msec on 17 subjects, this value *could be* obtained by *freely assigning* any values you wish to 16 subjects. However, you would then be *forced* to give the seventeenth subject the particular value required for obtaining the given mean.) Thus, since we had 17 subjects for each treatment, there were 16 + 16 = 32 degrees of freedom or *df*.

There are no values listed in the table for exactly 32 *df* but those for 30 *df* will do. We can see that there is very little difference between required values of t for 30 and 35 *df*. At the .05 alpha level a t of 2.04 is required to reject the null hypothesis; at the .01 alpha level a t of 2.75 is required. The value of t, 3.77, for our experiment clearly shows that our obtained difference of 23 falls in a *rejection region*, even using the .01 alpha level. The probabilities are shown as in Figure 6.1(c). Our statistical decision will be to *reject* the null hypothesis.

The distribution, when scaled off in t values, is a sampling distribution of t. The exact shape of the t distribution will differ according to the number of degrees of freedom in the experiment. That is why you have to look up the critical values to find whether your difference is significant.

The Null Hypothesis and ω^2

In this statistical supplement, it has been seen that there is a strong effect of the independent variable in the reaction-time experiment, est $\omega^2 = .28$. Here we have seen that it is most improbable that we would have obtained as large a difference as we did, if the null hypothesis were true. Do not get these concepts—strength of effect and statistical significance—confused. With very reliable data, even a small difference between means will allow rejection of the null hypothesis. Thus there can be a statistically significant difference with only a weak effect of the independent variable.

PROBLEM: Compute t and test the null hypothesis at the .01 alpha level for a reaction time experiment comparing choice between two lights (Treatment C) and choice between two tones (Treatment D).

Treatment C (Lights)				Treatment D (Tones)			
Subjects	RT	Subject	RT	Subjects	RT	Subject	RT
1	304	10	275	1	272	10	261
2	268	11	268	2	264	11	250
3	272	12	254	3	256	12	228
4	262	13	245	4	269	13	257
5	283	14	253	5	285	14	214
6	265	15	235	6	247	15	242
7	286	16	260	7	250	16	222
8	257	17	246	8	245	17	234
9	279			9	251		

Answer:

$$M_C = 265; \quad M_D = 250$$

$$s_C^2 = 292 \quad s_D^2 = 337$$

$$t = 2.47$$

The null hypothesis can be rejected at the .05 alpha level but not at the .01 alpha level.

STATISTICAL TABLE 2
Critical Values of *t* for Rejecting the Null Hypothesis[a]

DF	.05	.01
1	12.71	63.66
2	4.30	9.92
3	3.18	5.84
4	2.78	4.60
5	2.57	4.03
6	2.45	3.71
7	2.36	3.50
8	2.31	3.36
9	2.26	3.25
10	2.23	3.17
11	2.20	3.11
12	2.18	3.06
13	2.16	3.01
14	2.14	2.98
15	2.13	2.95
16	2.12	2.92
17	2.11	2.90
18	2.10	2.88
19	2.09	2.86
20	2.09	2.84
21	2.08	2.83
22	2.07	2.82
23	2.07	2.81
24	2.06	2.80
25	2.06	2.79
26	2.06	2.78
27	2.05	2.77
28	2.05	2.76
29	2.04	2.76
30	2.04	2.75
35	2.03	2.72
40	2.02	2.71
45	2.02	2.69
50	2.01	2.68
60	2.00	2.66
70	2.00	2.65
80	1.99	2.64
90	1.99	2.63
100	1.98	2.63
120	1.98	2.62
150	1.98	2.61
200	1.97	2.60
300	1.97	2.59
400	1.97	2.59
500	1.96	2.59
1000	1.96	2.58
∞	1.96	2.58

[a]Statistical Table 2 is taken from Table IV of Fisher and Yates: *Statistical Tables for Biological, Agricultural and Medical Research*, published by Longman Group Ltd., London. (previously published by Oliver & Boyd, Edinburgh), and by permission of the authors and publishers.

MULTILEVEL
EXPERIMENTS

Let us suppose that three investigators, named *Ae*, *Bee*, and *Cie*, each wanted to find the most effective way of using a fixed amount of study time to memorize a list of items. Ae hypothesized that it would be best to dwell on each item a long time and to go through the list slowly only a few times, with a *long interval* of time between one item and the next on each presentation of the list. Bee hypothesized just the opposite relation: that it was best to go over the list quickly many times, with a very *short interval* of time between one item and the next on each presentation of the list. Cie held that it would not make much difference how the time was divided; all that would count would be total study time.

Let us further stretch our imaginations by supposing that Ae, Bee, and Cie individually decided to experiment on 16-item lists in which the subject is required to memorize two-digit numbers, each of which is paired with a different three-letter group (called a *trigram*). That is, the subject might have to learn to say "27" when he is shown *BAW* and "84" when shown *NUV*, and so on for the 16 such pairs. Total study time is set at 320 seconds. The procedure is to present the lists item by item according to the plan of how the study time is divided and then, after the 320 seconds (sec) of study time have been used up, to test the subject to find out how many items have been memorized.

Here is what Ae did and what he found. He used two treatments: (A) 1 sec between items (making 20 presentations of the list), and (B) 4 sec between items (making five presentations of the list). Sure enough, his hypothesis that the *longer* interval would be more effective was supported. The subjects with the 1-sec interval averaged about seven correct responses. The subjects with the 4-sec interval averaged about 13 correct responses.

Bee again used the 4-sec interval as one treatment (A) but compared it with the 20-sec interval (B) in which the list was presented only once. He also found about 13 correct responses for the 4-sec. interval but fewer than 10 for the 20-sec interval. He concluded that *his* hypothesis was supported; the *shorter* interval is more effective.

Cie used a 3-sec (A) and a 10-sec (B) interval. With both he found about 12 correct responses on the average. Again an hypothesis was supported; this time, that it doesn't matter how study time is divided.

Let us finally suppose you were the editor of the journal to which Ae, Bee, and Cie submitted their articles (simultaneously). You would be in a position to put the three sets of results together and to show the relation between the independent variable (interval between items) and the dependent variable (mean number correct) as is shown in Figure 7.1.

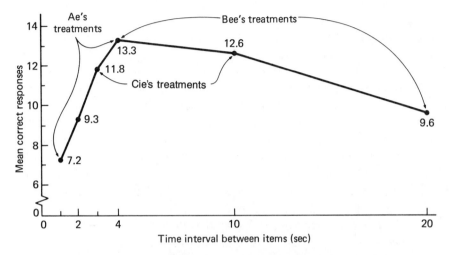

Figure 7.1 Results of the Calfee and Anderson experiment, showing the imaginary two-treatment experiments by Ae, Bee, and Cie. From Calfee and Anderson (1971, p. 242). Copyright 1971 by the American Psychological Association. Reprinted by permission.

Of course, you cannot really patch together an experiment in this way because of the many differences that must exist in the different studies. Figure 7.1 was drawn from actual data, but they did not come about piecemeal as was pretended. They are part of an extensive study on presentation rate effects by Robert Calfee and Rita Anderson (1971). The efforts of Ae, Bee, and Cie, also indicated in the graph, are all examples of two-treatment experiments (sometimes called *bivalent*). The experiments we have discussed thus far have been of this kind. The Calfee and Anderson study is a multilevel experiment (sometimes called *multivalent*); many *levels* of the independent variable were used. In fact, there were six levels of time interval between items: 1, 2, 3, 4, 10, and 20 sec.

As you have seen, there are advantages in going beyond a two-treatment experiment when it is possible to do so. It turned out that we knew very little from each of the separate experiments of Ae, Bee, and Cie. It was only by using a number of levels of the independent variable that we could get a real grasp of its relation with the dependent variable. In this chapter, we will go on to examine the advantages in some detail. After that, we will consider the experimental designs that can be used in multi-level experiments. We will find that our previous designs, those using between-groups controls and those using within-subject controls, are applicable. However, a new basic design is often more practical, the use of *across-subjects* controls. We will then go on to problems of internal validity in multilevel experiments.

What you should take with you from this chapter is a new way of looking at an experiment, as a means of getting at the relation between two *continuous* variables: As the independent variable changes bit by bit, how does the dependent variable change? You will look at two- or three-level experiments in the literature and try to see whether they could be converted to multilevel experiments. You will consider what the results might look like when this can be done. It is hoped that you will become aware of the dangers to internal validity in the typical designs used in multilevel experiments; and, in reading articles, that you will be able to judge whether the experimenter shared your concern.

The questions you will be expected to answer at the end of the chapter fall into these areas:

1. The control functions of multilevel experiments.
2. The more ambitious experimental hypotheses that can be tested only with multilevel experiments.
3. Different experimental designs that use across-subjects counterbalancing.
4. The issues concerning internal validity in choosing an experimental design for a multilevel experiment.

THE MULTILEVEL EXPERIMENT AS A CONTROL

We shall first consider what is accomplished by using a multilevel experiment when the experimental hypothesis could also be tested using only two treatments. An example is the experiment on the work ethic discussed in Chapter 5, in which the preference by Indian girls for active effort rather than freeloading was found. Two conditions were compared: (1) no bar pressing to receive a marble and (2) 10 lever presses to receive each marble. It would have been possible to use several amounts of active effort: e.g., 2, 5, 10, 50 bar presses for each marble. Would the experiment really be improved sufficiently to be worth testing at more levels? (Of course, the multilevel experiment should be preferred by the investigators since it would require more active effort on their part).

Less Likely To Miss an Effect

To see the advantage of the five-level experiment, we should first ask how the experimenters knew that each tenth lever press should be used for one of the two experimental conditions. It might have been too little work to make any real difference; it might have been too much. Apparently the investigators, Singh and Query (1971) believed that one lever press per marble would not have been enough. Also, at some very high level of number of lever presses per marble the girls would certainly have preferred to freeload, and before that point was reached there would have been a level equally attractive to freeloading. If Singh and Query had happened to use the equally attractive level, they would have been in the same position as imaginary experimenter Cie, with results suggesting that the independent variable had no effect. What this discussion has shown is that the hypothesis that was tested in the Singh and Query study was really a *quantitative* hypothesis relating amount of required lever pressing to amount of preference over freeloading. Such an hypothesis is *perfectly* tested only with a completely continuous independent variable. This is impossible, as it implies an *infinite* number of levels, separated by infinitesimal differences. Still, if five levels were used, the entire relationship between the independent and dependent variables would be approached. As the number of levels is reduced, the danger of *misrepresenting* this relationship increases. Hence, it can be said that the *internal validity* is better when such an hypothesis is tested with five levels as compared with an experiment with only two levels of the independent variable. This threat to internal validity arises from *incompleteness* of the independent variable. The previous threats described arose from unreliability or confounding with other variables, either procedural or associative (see Chapter 5, page 158). In the imaginary experiments of Ae, Bee, and Cie, we had a dramatic example of how the relation between the independent and dependent variables might be misrepresented through use of only a few levels. In addition, the two-treatment experiment has a further problem in respect to associative confounding, as will be seen in the following section.

Better Control over Associative Confounding

Evidence of the effect of an independent variable is not convincing if there are clear possibilities for associative confounding, the active level of the independent variable being associated with the active level of another variable. Control conditions, such as those described in Chapter 5, might not be as thoroughly effective as they seem to be.

Suppose an experiment is conducted, using college students as subjects, to find whether the drug caffeine acts on the nervous system to increase responsiveness to stimulation. The response measured is reaction time. There would be a clear danger of such confounding if only two treatments were used. Let us say that the active level has been set at 3 milligrams of the drug for each kilogram of the subject's weight. (On the basis of other studies, this is probably a good choice.) As was discussed in Chapter 5, the comparison treatment should be a *placebo*. If the caffeine is taken in the form of a pill, the placebo would be a pill with no active substance. Still, the placebo may not be a sufficient control for the subject's *awareness* of treatment. What is overlooked is the fact that a subject (especially a bright college student) can often tell whether the pill was caffeine or placebo. With the former, there may be a slight tingling in the fingers, a warmth of the face, quicker breathing, etc. This awareness could very well affect reaction time performance indirectly. The subject might expect to respond quickly and become more tense, thus shortening reaction time. Such an effect of caffeine has nothing to do with the experimenter's hypothesis of a direct influence on the nervous system. There has been associative confounding. The active level of the independent variable (3 milligrams/kilogram) is associated with the active level of a second variable, awareness of drug.

If, instead of a two-treatment experiment, a multilevel experiment is performed with dosages of 0, 1, 2, 3, 4, and 5 milligrams/kilogram of caffeine, a curve might be obtained as shown in Figure 7.2. The zero-dosage level, has, of course, employed a placebo. It is seen that there is a regular shortening of reaction time from Dosage 1 through Dosage 5. The sharper drop from 0 to 1 could, in part, be a drug awareness effect. However, the regular change thereafter provides rather convincing evidence that caffeine does increase responsiveness directly. It is unlikely that more and more awareness from stronger and stronger dosages would produce this curve. Still another treatment could be used if the experimenter is curious about the placebo effect. This would be to give no pill at all. The triangle in Figure 7.2 represents this treatment. The shorter

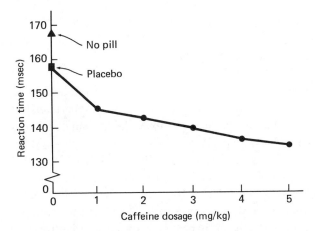

Figure 7.2 Results of an imaginary multilevel experiment to find the effect of caffeine on reaction time.

reaction time with the zero (placebo) dosage shows that there was indeed a placebo effect, as well as a drug awareness effect, as well as a true effect of caffeine. A lot can go on in a simple experiment!

There is often danger of such associative confounding in two-treatment experiments. In the experiment on the work ethic when a girl pressed the lever she was perhaps doing more than earning 1/10 of a marble. She was also making contact with the hidden experimenter (or whatever she supposed was in the big box). To some extent, then, the lever-pressing treatment might be associated with the active level of a second variable, making contact with a hidden experimenter. In a multilevel experiment using different amounts of required lever presses, a gradual increase in preference as the number of required presses was increased would be clearer evidence in favor of the hypothesis. However, if even a slight amount of work (e.g., a marble for each press) was preferred as much as a greater amount of work (e.g., a marble for each five presses), the evidence for the hypothesis would be weak. Again, a multilevel experiment provides a safeguard against associative confounding.

QUANTITATIVE AND QUALITATIVE INDEPENDENT VARIABLES

The multilevel experiment provides advances over two-treatment experiments that go beyond the controls just described. In general, it may be said that it furthers the goal of understanding by allowing the testing of more detailed experimental hypotheses. These advances will be described in this section and in the three that follow.

First of all, there is an advance that comes about, not because a number of levels of the independent variable are used, but simply because the independent variable is *quantitative*. The comparison here is with *qualitative* independent variables—those which cannot be described by means of numbers. An experiment need not be bivalent to have a qualitative independent variable. We saw in Chapter 4 that Gatewood and Perloff (1973) compared three methods of providing unit price information to supermarket shoppers. The differences between the three treatments were qualitative, because they cannot be described by numbers. On the other hand, even though the experiment on the work ethic was bivalent (just two treatments of the independent variable), the independent variable was basically quantitative, as was discussed.

The mere ability to devise a quantitative or graded independent variable demonstrates progress in *isolating* an effective factor. Let us compare similar experiments, one in which a qualitative independent variable is used and the other in which the independent variable is quantitative. An experimenter might hypothesize that symbolic information is processed more quickly by vision than by hearing. He then does an experiment comparing reading and listening. With the text material he has selected, he finds that his subjects can read, with good comprehension, at the rate of about 1,000 words per minute; however, equal comprehension of the same material cannot be attained by listening when the speaker exceeds 200 words per minute. There is no doubt that he has shown that a person can read more quickly than he can listen. However, it is another thing to say that this result supports the hypothesis of quicker processing of symbolic information by vision.

Reading and listening differ in many ways. One important way is that in reading the person may look ahead or look back for key words or ideas. In listening, the person has to take the word sounds one at a time and rely on his expectations or memory for expanding his "present time."

The experimenter decides to follow this lead to learn whether reading can be made worse by limiting the amount the reader can see at one time. He does this by moving each line of print past an open viewing window. The independent variable is the width of the viewing window— whether it shows, on the average, one word, two words, three words, five words, or ten words. It would probably happen that speed of reading with comprehension would increase as the window was made wider (up to a point). As a result of this experiment with a quantitative independent variable, something rather definite would be learned about the factors that control speed of reading. We would also be told to be careful in attributing the difference between reading and listening to any basic difference between speed of visual and auditory processing.

To be sure, the comparison between reading speed and listening speed is a valid experiment if that is the extent of the experimental hypothesis. It is not an hypothesis that adds greatly to understanding, because the independent variable is a "package." Testing the experimental hypothesis in the viewing window experiment would be more informative because the different treatments were different amounts of a better isolated variable. We would have better understanding from it of the reason for a difference in behavior.

As we have seen by now, there are many opportunities for devising quantitative independent variables: dosage of a drug, time interval between items on a list, number of items held in memory, required number of lever presses, varied weights to be lifted, etc. *Behavioral* scaling may also be used. For example, the trigrams of Calfee and Anderson were of "high meaningfulness." A large group of earlier subjects had rated many trigrams on meaningfulness. Those which were then used by Calfee and Anderson had all been given a high rating.

Verbal material has also been scaled on such diverse characteristics as *pronounceability* and *emotionality*. *Association* value is more often found in terms of performance—how many associations subjects can give to the item within a fixed period of time. Tables have been published with many kinds of scaled items (e.g., Runquist, 1966). If an experimenter wishes to, he can thus use as an independent variable in an experiment on meaningfulness, pronounceability, emotionality, or association value. For example, he may investigate the relation between pronounceability and speed of memorizing.

It is also possible for an experiment to scale larger units of verbal material. For example, sentences may be scaled by ratings as to pleasantness or as to clarity. The funniness of jokes may even be scaled. Of course, there is the problem that subjects might not agree. This is best handled by starting with a large number of items—say, 200 jokes. Fifty judges could then rate them from 10 (screamingly funny) down to 1 (blah). The jokes would be kept in the scale only if there was very good agreement among judges. This might leave only 60 or so of the original jokes. The scale value of each joke would simply be its average rating. Thus there might be one with a high average rating, 9.4; one with a moderate rating, 5.3; one with a low rating, 1.5. With such material, an experimental hypothesis might be tested that one will remember very funny jokes and really painful ones rather than moderately funny jokes. Of course, the original raters could not be used as subjects.

THE HYPOTHESIS OF A MAXIMUM (OR MINIMUM) VALUE

An Experiment on Presentation Rate and Memorizing

Let us now return to the experiment by Calfee and Anderson (1971), in which total study time for memorizing a list was divided in different ways. As can be seen in Figure 7.1, the six different time intervals between items on a list made up the six levels of that independent variable. A separate group of 20 subjects was tested at each interval. A *maximum* value of the dependent variable—slightly over 13 mean correct responses —occurred with the 4-sec interval, with five presentations of the 16-item list accounting for the 320 seconds of study time. This finding was just about what Calfee and Anderson had expected. We will shortly go into the reasons why an experimental hypothesis of a maximum value for one of the intermediate levels of the independent variable made sense.

An Experiment on Shock and Learning

If Calfee and Anderson were not surprised by their results, two *much* earlier investigators (1908!) certainly were. At this point in time, some 70 years later, we may snidely second-guess these pioneers and tell them what their hypothesis should have been. Let us first get down to the experiment.

Robert Yerkes and John Dodson were the experimenters. Yerkes, who was working his way up the evolutionary scale in his research, was then exhaustively studying a curious little creature called the *dancing mouse* (1907). Because of a genetic deficiency, this inbred strain of house mouse occasionally spins around in circles and figure eights. In later years, Yerkes culminated his career with studies of chimpanzees and army draftees.

The task given to a mouse (1908) was to learn a black-white discrimination habit. There were doors to two tunnels side by side. On a given trial, a mouse was gently forced to enter one or the other tunnel. This was done by gradually reducing the space in the entrance chamber to the two tunnels by means of a sheet of cardboard. One of the tunnels had white cardboard for sides and ceiling as well as the area around the entrance. The other was similarly fitted out with black cardboard. Electric grids formed the floor of each tunnel. However, the current was turned on only when the mouse entered the *white* tunnel. In either case, after it ran through a tunnel it had a path to run back to its nest box. There, waiting for it was a dancing mouse of the opposite sex. (What was lacking in experimental control was made up for in compassion.)

The mouse could not learn to avoid the shock by simply running to the left or to the right on each trial since the two tunnels were made light or dark randomly from trial to trial. Each animal tested was given 10 trials a day. This was continued until it made all of its runs on three days in a row without error. For example, Male No. 128 finally achieved this goal on Days 16, 17, and 18. Obviously it had mastered the discrimination at the end of Day 15. Hence, this animal was stated to have met the *criterion* of learning in 150 trials, since there were 10 trials per day.

Different· strengths of shock were used for different mice. They were given in "units of stimulation." The *weak* level [125 units] seemed barely noticed by a mouse. "The *strong* stimulus [500 units] was decidedly disagreeable to the experimenters and the mice reacted to it vigorously" (pp. 467–468). The *medium* level was 300 units. Four mice were tested at each strength of shock, two females and two males.

The dependent variable was the mean number of trials to reach the criterion by the four mice of a group. Results are shown in Figure 7.3. It

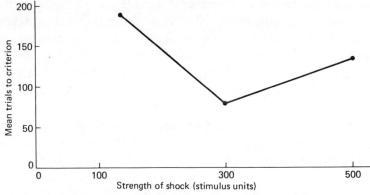

Figure 7.3 Relation between strength of shock and number of trials required by dancing mice for learning a black-white discrimination.

is seen that the *minimum* value of the dependent variable, 80 trials, was obtained at the *intermediate* level of the independent variable, 300 stimulation units. As a matter of fact, learning with the weakest stimulation was even worse than was shown, since one of the four mice never did learn the discrimination and its score was arbitrarily set at 200 trials, the point at which the tests were discontinued.

Of course, the use of only three levels of the independent variable makes the experiment barely qualify as multilevel. Yerkes and Dodson were unhappy about this too: "Attention should be called to the fact that since only three strengths of stimulus were used . . . it is possible that the most favorable strength of stimulation was not discovered" (1908, p. 482).

The Advance in Understanding

There are many experiments in which the hypothesis of a maximum or minimum is reasonable. By the way, one should not take the terms *maximum* and *minimum* to stand for different ideas in these experiments. If the Yerkes and Dodson criterion measure had been number of correct responses, a *maximum* would have been found at 300 stimulation units. If Calfee and Anderson had used mean number of errors rather than of correct responses as their dependent variable, there would have been a *minimum* value at the 4-sec interval.

It will be recalled that the quantitative hypothesis for Singh and Query's experiment on the work ethic was that there would be increasing preference as more lever presses were required—but *only up to a point*. That is, after the level was obtained to provide the maximum, further increase would reduce the preference. The reasoning for that expectation was that two things happen as the number of required lever pressings was increased, assuming with the investigators that their effect was due to a feeling of "acting on the world." At first, there would be gradual gain in "acting on the world" as greater effort was required to obtain each marble. That would increase the attractiveness of lever pressing. However, effort cannot be entirely pleasant, and as the number of presses was further increased there eventually must be a greater and greater "noxious" effect. The peak of maximum preference for lever pressing over freeloading would come at that level of lever pressing in which the difference between the "acting on the world" effect and the "noxious" effect is largest. One reason, then, for the hypothesis of a maximum (or minimum) is the theory of two opposing underlying processes determined by the independent variable, with the "negative" process becoming stronger than the "positive" process as high levels of the independent variable are reached.

Calfee and Anderson (1971) have identified a different kind of opposition for the experiment on presentation rate and memory. In order to say the right numbers when shown a trigram, the subject must first of all not get these trigrams mixed up. This is called *stimulus differentiation*. It is improved by a *short* interval between items. Second, the subject must learn to connect each number with its paired trigram. This is called the *associative process*. It is helped by a *long* time between items. There must be some time interval that is optimal. For any shorter interval, the gain in stimulus differentiation is more than offset by the loss in the associative process. For any longer interval, the gain in the associative process is more than offset by the loss in stimulus differentiation. Thus, a second basis for expecting a maximum (or minimum) is a theory that increases in the independent variable bring opposite changes in two underlying processes, both of which are positive. The maximum or minimum occurs at the level that affords the optimal combination of the two processes.

It may be argued that a dancing mouse is in very much the same position in learning to avoid the electric shock in the white tunnel. It must distinguish between the two tunnels, and it must associate a tunnel with shock or nonshock. A statement by Yerkes and Dodson (1908, p. 476) indicates that differentiation was poor with too strong a shock: "The behavior of the dancers varied with the strength of the stimulus to which they were subjected. They chose no less quickly in the case of the strong stimuli than in the case of the weak, but they were less careful in the former case and chose with less deliberation and certainty." Stimulus differentiation, then, becomes poorer as shock is increased. Association of white tunnel with shock (provided that differentiation has occurred) can only become more impressive as the strength of the shock is increased. Again, there must be some level of the independent variable, strength of shock, which is optimal for discrimination learning.

There must be some of you who have gone onto the next stage of reasoning. It is that a more difficult discrimination should require more deliberation than an easy one. First of all, of course, this would mean that it is learned more slowly. Moreover, it suggests that it should be learned optimally with a weaker shock than would be the case for the easy discrimination. In subsequent work described in their article, that is exactly what the investigators found. This is their statement (1908, p. 481): "As the difficultness of the discrimination is increased, the strength of that stimulus which is most favorable to habit formation approaches the threshold." That relation is known today as the *Yerkes-Dodson law*. But we are getting ahead of ourselves and will return to these experiments in the next chapter.

What has been shown for the three experiments described is that a multilevel experiment can test a hypothesis concerning two underlying processes in which there is an aspect of opposition related to level. The actual experimental hypothesis is that there will be a maximum (or minimum) value of the dependent variable for some intermediate level of the independent variable.

HYPOTHESES OF ABSOLUTE AND PROPORTIONAL RELATIONS

In the three experiments to be discussed in this section, the experimental hypothesis was that the dependent variable would change gradually as the independent variable changed gradually. Yet the hypothesized relation was of a somewhat different kind in the three cases. We shall start with the simplest relation and proceed to the most complex (but not all *that* complex). We may think of a change in either *absolute* or *proportional* terms. Thus the increase from 4 to 6 may be described either as an absolute increase of 2 or as a proportional increase of .5 (i.e., 6 is 50 percent higher than 4).

In one of the three experiments, it is hypothesized that equal *absolute* changes in the independent variable will result in equal *absolute* changes in the dependent variable. In the second experiment, it is hypothesized that equal *proportional* changes in the independent variable will result in equal *absolute* changes in the dependent variable. In the third experiment, it is hypothesized that equal *proportional* changes in the independent variable will result in equal *proportional* changes in the dependent variable. This is beginning to sound pretty abstract, so let us get on to the experiments themselves.

An Experiment with an Absolute-Absolute Hypothesis: Memory Search

Let us consider an experiment in which a digit is suddenly shown to a human subject. If it is one of two digits (say, 2 or 5), the subject presses a right-hand button; if it is not part of that set (0, 1, 3, 4, 6, 7, 8, or 9) he presses the left-hand button. Reaction time is found from the instant the "probe" digit appears until a response button is pressed. Other sizes of the *positive set* are used on different blocks of trials: 1, 3, 4, 5, or 6 digits. What is found is the mean reaction time for each of the sizes of the positive set. Then the line is drawn relating reaction time to set size. The results of the experiment (Sternberg, 1969) are shown in Figure 7.4. As can be seen, a line connecting all of the points would not be perfectly straight. However, the slight waviness is most probably due to some unsystematic variation. We may state that in order to obtain a uniform absolute increase in reaction time (of 35 msec), the level of the independent variable, set size, must be increased by one item (again, an absolute term).

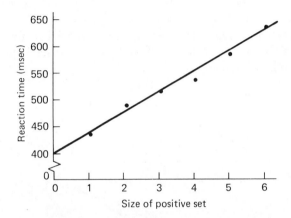

Figure 7.4 Relation between size of positive set and reaction time to a probe item (Sternberg's experiment). From Sternberg (1972, p. 292). Copyright 1969 by North Holland Publishing Company. Reproduced by permission.

An Experiment with a Proportional-Absolute Hypothesis: Choice Response

Two experimenters, W. E. Hick in England and Ray Hyman in the United States, did similar experiments, at about the same time, within the theoretical framework called *information theory* (Hick, 1952; Hyman, 1953). They found a different lawful relation between number of alternatives and reaction time than did Sternberg for set size. Hick's procedure is a little easier to explain than Hyman's, so it will be used as the example.

There was a half circle of 10 small electric lamps before the subject. His fingers (including the thumbs) rested lightly on 10 telegraph keys. When a light went on, the corresponding key was to be pressed. Ten alternatives was the highest level tested of the independent variable. There were also the conditions in which 8 or 6 or 5 or 4 or 3 or 2 or just one of the 10 lamps could go on. Each new signal occurred 5 sec after the previous response.

Separate blocks of trials were run with either 1, 2, 3, 4, 5, 6, 8, or 10 alternatives. Mean reaction time was found for each of these levels. The line relating reaction time to number of alternatives is shown in Figure 7.5 in two different ways. In (a) the horizontal scale for the independent variable is absolute, just as in the case for the Sternberg data. Each increase in number of alternatives is given the same distance. However, here the line is not straight but curves downward. Each new unit of choice brings about a smaller increase in reaction time rather than the same amount. When the horizontal scale is changed to that of the right-hand graph in (b), the line straightens out.

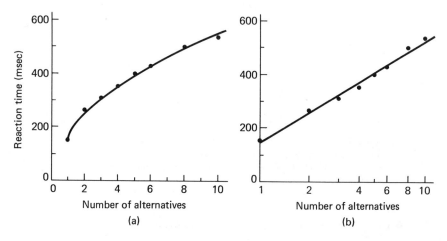

Figure 7.5 Relation between number of alternatives and reaction time (Hick's experiment). From Hick (1952, p. 15). Copyright 1952 by Academic Press, Inc. Reproduced by permission.

In (b) the horizontal scale is marked off with proportional spacing. (The technical name for such a scale is *logarithmic*.) Now, for example, each doubling of the number of alternatives—from 1 to 2, from 2 to 4, from 3 to 6—takes up the same distance. Go ahead; measure the scale to make sure. Thus, the straight line showing the relation between the number of alternatives and reaction time gives support to the proportional-absolute hypothesis. As the number of alternatives is increased by an equal proportion there is equal absolute lengthening of reaction time. The doubling of the number of alternatives is a proportional increase of 1 (i.e., of 100 percent). Each such doubling increases reaction time by 110 msec, the same absolute amount.

An Experiment with a Proportional-Proportional Hypothesis: Subjective Heaviness

Does a 1-pound weight *feel* half as heavy as a 2-pound weight? That is the kind of question asked by S. S. Stevens in a series of experiments conducted over many years, using not only weights but also tones, lights, odors, etc. Robert Harper and Stevens went about an experiment on weights in a very direct manner (1948). The subject stood before a table on which there were seven identical closed containers. One was put aside and called the "standard." The task was to lift the standard and each of the six other weights, and then to pick out the one "which feels half as heavy as the standard" (p. 344). Please note: The subjects were *not* to try to figure out which one actually weighed half as much.

It was found that the weight that felt half as heavy as a 100-gram standard was—on the average—a weight of 72 grams. The experimenters put this relation by numbers thus: A weight of 100 grams has a subjective heaviness of 1 "veg" (just by definition), so a weight of 72 grams has a subjective heaviness of ½ veg. In case you are wondering, Harper and Stevens got the word *veg* from an Old Norse word that means "to lift" (p. 345).

On other trials, a weight different than 100 grams was used as the standard. It happened that when the *standard* weight was 140 grams, the weight that felt half as heavy was—on the average—100 grams. Since 100 grams is defined as having a subjective heaviness of 1 veg, 140 grams, feeling twice as heavy—has a subjective heaviness of 2 vegs. In all, there were eight such trials, with the standard being as low as 20 grams and as high as 2,000 grams. When all the information was put together, the smoothed curve relating weight lifted to its subjective heaviness was that shown in Figure 7.6(a).

As can be seen, when the horizontal and vertical axes are both scaled off in absolute units (e.g., the distance between 100 and 200 grams

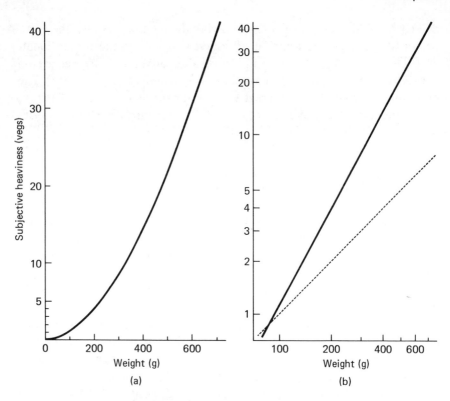

Figure 7.6 Relation between the experienced sensation of heaviness with increase in lifted weight (Harper and Stevens experiment). Reproduced by permission of the *American Journal of Psychology*, Harper and Stevens, and The University of Illinois Press (1948, p. 346). Copyright 1948 by Karl M. Dallenbach.

is the same as that between 500 and 600 grams; the distance between 10 and 20 vegs is the same as that between 30 and 40 vegs) the line is not straight, but curves *up*. However, the data are fairly well fit by the straight line in (b), where *both* axes are now scaled off in proportional units (e.g., the distance between 100 and 200 grams is the same as that between 400 and 800 grams; the distance between 2 and 4 vegs is the same as that between 8 and 16 vegs).

Thus, a proportional-proportional hypothesis was supported. Whenever you increase a weight by a given proportion, you increase the subjective heaviness by a constant proportion. It should also be noted that the axes were scaled off in *equal* proportional units. For example, the distance between 2 and 4 vegs is the same as that between 100 and 200 grams. If a proportional increase in weight brings about the *same* proportional increase in subjective heaviness, the plotted line would rise at a 45° angle (shown by the dotted line). In this experiment a proportional increase of one (i.e., 100 percent) of the lifted weight brings about the greater proportional increase of about two and a half (i.e., 250 percent) in subjective heaviness. Put in other words, doubling the lifted weight almost quadruples the subjective heaviness. Getting back to the original question, you can see that a 1-pound weight does *not* feel half as heavy as a 2-pound weight; it feels considerably *less* heavy than that.

The Advance in Understanding

Each of the three straight-line relations reflects a theory or model of processes underlying behavior. The basis of Sternberg's absolute-absolute hypothesis is as follows. Each of the digits in the positive set is held in memory by the subject. When a probe digit appears, the items in the positive set are "scanned" in serial order to determine whether the probe digit was one of those in the positive set. In other words, the probe is compared with each item in the positive set in turn. If each such comparison requires 35 msec, that will make the total scanning time just that much longer when an additional item is added to the positive set. Thus the fact that the same absolute increase was found with each absolute increase of one item was taken as support for the serial scanning model.

Hick's proportional-absolute hypothesis of the relation of reaction time to the number of alternate choices follows from a different kind of model of underlying processes. The idea here is that the subject performs a choice response task in the most strategic way: by a series of simple decisions. Thus, if there are eight alternatives, the first simple decision is between Alternative Group 1, 2, 3, 4 and Alternative Group 5, 6, 7, 8. Suppose the correct choice is Alternative 7. In that case, the first decision should be for Group 5, 6, 7, 8. The next simple decision is between Group 5 and 6 and Group 7 and 8. Thus, the correct second simple decision is Group 7 and 8. This leaves only the decision between 7 and 8. Thus, the third and last simple decision is Alternative 7. In all, there were only three simple decisions for eight alternatives. The following will thus hold for different numbers of alternatives: two alternatives, one decision; four alternatives, two decisions; eight alternatives, three decisions. If each simple decision takes the same amount of time, estimated here to be 110 msec, there should be the same absolute increase in reaction time (110 msec) for each proportional increase of one (100 percent) in the number of alternatives. Of course the theory must be made somewhat more elaborate, to take care of other numbers of alternatives that do not divide so nicely, e.g., six or ten.

Stevens' hypothesis of a proportional increase in experienced sensation with a proportional increase of the stimulus is based on his conception of the way in which a sense organ must convert the physical energy acting on it. A proportional increase in energy brings about a proportional increase in neural activity.

In each of these three examples, the hypothesis tested and the results obtained reflect a depth of understanding that goes far beyond the mere existence of a variable that affects behavior. Because of the exact relation between the independent and dependent variables (both conceived as continuous), we have gained insight into *how* the relation might come about.

PREVIOUS EXPERIMENTAL DESIGNS APPLIED TO MULTILEVEL EXPERIMENTS

Two basic experimental designs have been described in previous chapters. These are the between-groups design in which a different group of subjects is assigned to each treatment and the design using within-subject controls. These two designs will be considered here in relation to multilevel experiments. From this discussion, we will see why investigators have been attracted to the third basic design, which will be described in the following section.

Between-Groups Designs

Calfee and Anderson (1971) used a between-groups design in their experiment on memorizing. Six different intervals between items were used: 1, 2, 3, 4, 10, and 20 sec. A group of 20 subjects was assigned to each of these intervals, for a total of 120. (This was only one-fourth of a larger experiment, so 480 subjects were tested in all!) Subjects were assigned to levels randomly as they appeared. One disadvantage of the use of a between-groups design for a multilevel experiment is readily apparent. If a fairly large number of subjects is required at each level in order to equate the groups, the total number required may be so large as to be impractical.

Designs Using Within-Subject Controls

You will recall that in the first experiments described in this book the main threats to internal validity were from changes over time and from sequence effects, since the same subject was given the different treatments. Controls were necessarily within the same subject. The three orders of treatments that were used as the within-subject controls were: alternating (weavers), counterbalanced (piano pieces), and random (tomato juice). Such a design was used by Hick in his study of number of alternatives and reaction time: he was the only subject.

However, this design may be expanded if a number of subjects are to be used, as is typical. The same within-subject control will be used on each of the subjects. When a large number of brief trials are given in one block and when the subject must be kept in ignorance regarding the level on any given trial, a random order is typically used. For example, in one experiment (Gottsdanker and Way, 1966), the time interval between two signals for response was varied randomly from trial to trial. These were all short intervals: 50, 100, 200, 400, and 800 msec. The experimenters were testing the hypothesis that reaction time to the second signal would shorten evenly as interval was lengthened (up to a point). It was important that the subject not know what interval was coming next. In a block of 100 trials, each of the five levels occurred randomly 20 times. The same random order was used for each of eight subjects.

However, the use of within-subject controls does not work out as well in a multilevel experiment when a counterbalanced order is called for, as when each trial takes considerable time (e.g., the memorizing of a piano piece). The counterbalanced order for two treatments was symbolized as ABBA, where A and B represent a trial with one of the two treatments. For a multilevel experiment, the within-subject counterbalanced order for a six-level experiment would be ABCDEFFEDCBA. The problem is that if each trial is lengthy, it may be impractical to give a subject so many trials.

DESIGNS USING ACROSS-SUBJECTS COUNTERBALANCING

We have now seen that in a multilevel investigation a between-groups design may require too many subjects in order to equate groups and that a design using within-subject controls may require too much time per sub-

237

ject in order to control for sequence effects. A way out is to give each subject each treatment but to control for sequence effects *across* subjects. The same subjects are compared over all for the different levels, and each subject is tested only once at each level. Thus one subject (or group) could be given the treatments in order ABCDEF and another subject (or group) be given the order FEDCBA. Such designs, since each subject is given more than one treatment, are usually classed, along with those in which within-subject controls are used, as repeated-measures designs. However, the difference is important. When within-subject controls are used, the trials for each subject make up a complete experiment. As far as internal validity is concerned, a group of subjects is used in order to improve reliability, not to control for systematic confounding. When across-subjects controls are used, the results for any one subject are known to be tainted with systematic confounding. More than one subject is required for any control over systematic confounding. We will now describe the three most widely used designs for multilevel experiments using across-subjects counterbalancing.

Reverse Counterbalancing

Reverse counterbalancing is the design just commented on. It may be represented as follows:

Subject Group	Sequence
1	C B A D E (any order)
2	E D A B C (the reverse order)

What this means is that only two sequences of levels are used. As was shown, they need not be ABCDE and EDCBA, where A means the lowest level of the independent variable and E means the highest level. In another part of the Gottsdanker and Way experiment that was introduced earlier, time interval between the two signals for response remained fixed over a block of trials. (Subjects could not simply make both responses when the first signal occurred, since each response required a choice.) For one group of four subjects, five blocks of 100 trials were given with time intervals in the order 50, 100, 200, 400, and 800 msec. (i.e., ABCDE). For the other group of four subjects, the order was 800, 400, 200, 100, and 50 msec. (i.e., EDCBA).

What reverse counterbalancing does is to give each level the same *average* position over the two sequences. Thus, with the two orders shown, CBADE and EDABC, Level E is in Positions 5 and 1, for an average of 3. Level D is in Positions 4 and 2, again for an average of 3, etc. This is an effective control only if transfer effects are uniform. In other words, the assumption is made that there is the same carryover from Position 1 in the sequence to Position 2 as there is from 2 to 3, or from 3 to 4, or from 5 to 6.

A serious possibility is *nonuniform* transfer, described in Chapter 2 relative to within-subject designs. Suppose there are learning effects that improve performance uniformly up to the third trial in a sequence, but not beyond. For subjects having a sequence CBADE, the later levels—A, D, and E—would be favored equally. For subjects having the reverse sequence, EDABC, the later levels, A, B and C, would be favored equally. Thus Level A, which was in the middle of the two sequences would be most favored overall and C and E least favored. If there was carryover of fatigue rather than of learning, the level in the middle of the two sequences would be the most disfavored.

If the transfer effect is different at the various positional orders in the sequence, the confounding variable is *amount of transfer*. In the foregoing sequence, CBADE, the amount of transfer could be described as follows: 0 for C (since it was the first treatment), 1 for B, and 2 for A, D, and E (since transfer did not increase beyond the third trial). Similarly, for the reverse sequence, EDABC, the amounts would be: 0 for E, 1 for D, and 2 for A, B, and C. In sum total, the amounts of transfer would be: 4 for A, 3 for B and D, and 2 for C and E. Because of this inadequacy of reverse counterbalancing, investigators have turned to designs with better controls. These will now be described.

Complete Counterbalancing

In order to avoid the systematic confounding due to nonuniform transfer effects that occur with reverse counterbalancing, all possible sequences of the levels may be used rather than just two. This design, with complete counterbalancing, is represented for a three-level experiment as follows:

Subject Groups	Sequences
1	A B C
2	A C B
3	B A C
4	B C A
5	C A B
6	C B A

Thus, if in the Gottsdanker and Way study only three levels of the independent variable had been used (e.g., 50, 100, and 200 msec) the different subjects—or groups of subjects—would have had these six sequences: 50, 100, and 200 msec; 50, 200, and 100 msec; 100, 50, and 200 msec; 100, 200, and 50 msec; 200, 50, and 100 msec; 200, 100, and 50 msec. The reason that complete counterbalancing was not illustrated for more levels of the independent variable (as would usually be the case for a multilevel experiment) is that the table would have to be very large. All five levels of the Gottsdanker and the Way study would demand 120 sequences. Thus, even if only one subject was used per sequence, this would mean 120 subjects. The number of sequences required with complete counterbalancing is given by n factorial, where n is the number of levels. For six levels, n factorial is found by the following series of multiplications:

$$6 \times 5 \times 4 \times 3 \times 2 \times 1 = 720.$$

Since across-subjects counterbalancing was introduced to get by with *fewer* subjects than are required by a between-groups design, complete counterbalancing is seldom used. The next design is a way of reducing the number of subjects required while avoiding the assumption of uniform transfer that is basic to reverse counterbalancing.

Latin Square

The idea that would naturally occur, if one does not want to use all possible sequences, is to make a random selection among the dozens that are possible. This is, in fact, sometimes done. However, in a set of sequences chosen randomly it is unlikely that each level would fall equally often at each position in the sequence. Thus, the danger from nonuniform transfer would still be present.

What is done is to make a random selection among "squares" in which each level does appear *once* at each serial position. Each such square is a complete experimental design. It is called a *Latin square*. Here is one of the 8,640 such squares for six levels of an independent variable:

Subject Groups	Sequences
1	A B C D E F
2	C E D A F B
3	E C A F B D
4	B D F C A E
5	D F B E C A
6	F A E B D C

Since, in a Latin square, each level appears at each sequential position, there are naturally as many subject groups as there are levels of the independent variable. If Gottsdanker and Way had used a Latin square design rather than reverse counterbalancing (as they should have), they would have had five subject groups for their five levels of the independent variable. This would have meant use of either five or ten subjects rather than the eight they used. (Eight into five won't go.)

Investigators often make a restriction in the Latin square. That is, each level must be *preceded* once by each other level. Such a square is called a *balanced square*. In the Latin square just shown, this was not the case. For example, Level B was preceded immediately once by A and E, three times by F, and never by C and D. A method for obtaining balanced squares is given by Wagenaar (1969). An example is:

Subject Groups	Sequences
1	A B C D E F
2	B D A F C E
3	C A E B F D
4	D F B E A C
5	E C F A D B
6	F E D C B A

If all transfer effects were from the immediately preceding level, the balanced square would accomplish a great deal. Unfortunately, there is no way of knowing whether that is the case. Let us now consider the systematic sequential confoundings that can occur even with complete counterbalancing.

Range Effects

In a multilevel experiment, there is a range of levels of the independent variable from the lowest to the highest. With any design using counterbalancing, whether within-subject or across-subjects, present performance at a given level may differ according to whether the preceding levels were lower, higher, or mixed.

Asymmetrical Effects   These effects have already been commented on in Chapter 2 relative to within-subject designs. For example, previous experience with A may help B, but not vice versa. This idea may be enlarged on for multilevel experiments using across-subjects counterbalancing. Suppose there are five levels of the independent variable and complete counterbalancing has been used (all 120 sequences). While each level has been preceded once by every possible sequence of other levels, they have not been preceded by *identical* ones. Overall, lower levels have been preceded by higher levels, and vice versa. For example, there is no way in which the lowest level, A, could have been preceded by a series of levels containing a lower level. If, generally speaking, there is favorable positive transfer from lower to higher levels, but not vice versa, Level A will suffer the most. Thus, asymmetrical transfer in a multilevel experiment will favor or disfavor levels according to how close they are to one end or the other of the range of levels tested.

Centering Effects   Still another kind of range effect has been demonstrated in an experiment devised for that purpose by J. E. Kennedy and J. Landesman (1963). They performed the equivalent of two Latin square experiments with two groups of persons on a block-turning task. The range of levels on one experiment overlapped the range for the other. The independent variable was height of working surface. The dependent variable was the mean number of recessed blocks turned over on a 3-minute trial.

In Figure 7.7, the mean number of blocks turned is shown separately for the two groups. What is interesting is that the subjects in Condition A, in which the lowest level was −18 inches did their best performance

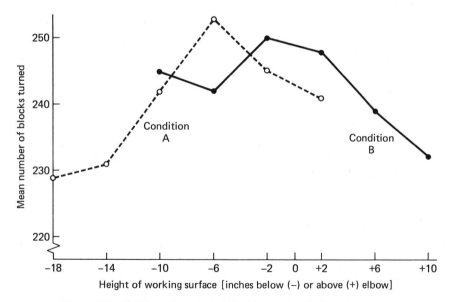

Figure 7.7 Relation of number of blocks turned to height of working surface for two overlapping ranges. From Kennedy and Landesman (1963, p. 203). Copyright 1963 by the American Psychological Association. Reprinted by permission.

at −6 inches, while the subjects in Condition B performed relatively poorly at that level. This latter group, whose lowest level was −10 inches performed best at −2 and at +2 inches.

In this experiment, levels near the *center* of the range of levels were favored, not those at one extreme or the other. The only levels that could have had both lower and higher levels on preceding trials in the sequences were, of course, those near the center. Possibly these middle levels seemed "typical" to the subjects and made them feel comfortable. However, your guess is as good as mine. What is clear is that a centering effect was found.

Complete counterbalancing and Latin square designs do not make the strong assumption of reverse counterbalancing that there is uniform transfer from position to position. However, the assumption is made that the relation between present and previous levels does not matter. There is much evidence to the contrary (Poulton, 1973). It is important whether the preceding levels tend to be lower, higher, or mixed.

What Can Be Done?

When across-subjects counterbalancing is used, it first of all would be a good idea to avoid reverse counterbalancing. Since complete counterbalancing is generally impractical this would mean that it would be a good idea to use a Latin square design, preferably a balanced square. Further, in order to avoid negative transfer effects because of fatigue, trials should be well spaced. It would also seem to be a good idea to divide the experiment in half and use overlapping ranges. If no range effects are then found, there is a good indication that sequential confounding has been avoided. As we shall see in the next section, across-subjects counterbalancing does have one major advantage over between-groups designs in multilevel experiments. The approach is too good a one to be abandoned because it can never produce a perfect experiment. What approach can?

CAN WE BELIEVE THE CURVES?

Use of a between-groups design will eliminate any possibility of the sequential effects just discussed. Each subject is tested at only one level. However, there are further threats to internal validity for those multilevel experiments that test hypotheses of the exact relation between independent and dependent variables that remain. We are talking about such experiments as that by Sternberg (1969), which tested an absolute-absolute hypothesis on the relation between size of positive set and time for memory search, and that by Hick (1952), which tested the hypothesis of a proportional-absolute relation between number of alternatives and reaction time. As a matter of fact, a between-groups design is more vulnerable to the first of these threats than is a design using counterbalancing.

Representation of the Individual

An imaginary set of data relating an independent and a dependent variable using between-groups comparisons is shown in Figure 7.8(a). Each small dot stands for the performance of a subject. The mean for each level is shown by a large dot, and the curve found is the line that connects the means. Let us now see how these data might have been obtained from an *ideal* experiment, a subject tested simultaneously at all levels.

In (b) is shown one way such data could come about with several such subjects. The numbers correspond to the same subject at the different levels. The lines connecting the performances of the "same subject" are seen to look very much like the line connecting the means in (a). Of course, another possibility is that the line through the means is not a good representation of the individual curves, as is shown in (c). When a between-groups experiment gives results presented in (a), there is no way of knowing which kind of picture, (b) or (c), is the correct one. Because of within-level variability among subjects, the shape of the curve is uncertain.

There are two ways of reducing this difficulty when using a between-groups design: matching subjects and use of homogeneous groups. If the subjects are given a pretest and then matched subjects at each level of performance are assigned to a different level of the experimental variable, the imaginary set of numbers shown in (b) or (c) can be made an actuality. The number 1 represents one matched set of subjects, 2 another matched set, and so on. We can see directly whether the consistent picture in (b) is correct or rather an erratic one such as that in (c).

The second method is by using only a very homogeneous group of subjects, again probably selected by a pretest. An example is shown in (d). Now it does not matter nearly as much how the lines between dots are drawn; the shape of the curve will be about the same. The two meth-

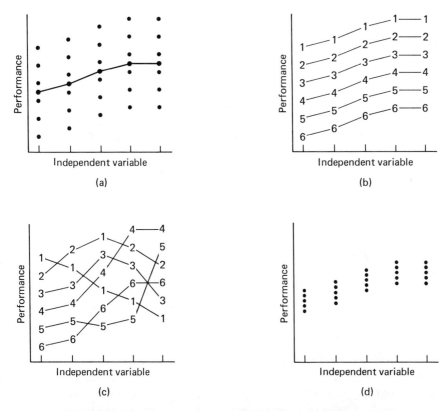

Figure 7.8 How an average curve could relate to data from an ideal experiment, each subject tested simultaneously at all levels: (a) the individual data and line of averages; (b) each subject's curve is the same as the average curve; (c) all subjects have different curves; and (d) homogeneous subjects—little chance for misrepresentation by the average curve.

ods can be combined by using only a homogeneous group and matching the individuals at the different levels of the independent variable.

At this point, you might be a bit concerned about the blatant bias in selecting certain individuals to be subjects and rejecting the rest. However, the meaningful conclusions about how behavior is related to level of the experimental variable are in no way compromised. Certainly they apply to a small segment of the population. However, that is a question of generality that is easily answered by testing other homogeneous groups, with higher and lower levels of performance.

If, instead, across-subjects counterbalancing had been used, with several subjects tested on each of the five sequences, a clearer picture would be available. Although the curve for any particular sequence is uncontrolled in respect to sequential confounding, there should be the *same* confounding for all subjects given that sequence. If, generally, the curves for subjects given the same sequence are all of the same shape, there is good evidence that the total group curve over all sequences does represent the individual. Because the same individual is tested at each level (although not simultaneously), the across-subjects design more closely approaches the ideal experiment—in this respect—than does the between-groups design. It has better internal validity in its representation of the individual.

Freedom from Distortion

If you were doing an experiment to find exactly how the accuracy of dart throwing is affected by weight of dart, you would want to make sure your measurements were not distorted. If you were using a foot ruler to measure how far the bullseye was missed on each throw, you would not want it to have three times as much space between the 9- and 10-inch marks on your ruler as between the 2- and 3-inch marks. (If it did, you would probably return the ruler to the trick-and-joke store.) Neither would you use a scale to measure the weight that hardly budged when you put a light dart on it but that went near to its maximum value with a barely heavier weight. You would know that, because of using these *distorted* measuring devices, the curve relating the independent variable (weight of dart) with the dependent variable (amount of error) would be an inaccurate representation. To be sure, you might find that, generally speaking, throwing became more accurate as weight increased. But you would not be able to test the hypothesis that the relation is an absolute-absolute one (e.g., 2 inches less error for each weight increase of 1 ounce).

Of course, you are not going to make this kind of mistake in an experiment like that just described. However, there are two kinds of measurement in which all care must be taken to avoid distortion. The first is that in which response is measured through very small physical changes. An example would be the recording of the galvanic skin response—change in resistance to electrical flow that occurs when a person is startled or tells a lie. In order for the response to be recorded, an electrical change must be amplified. How can we be sure that twice the movement of the recording pen means twice as large a galvanic skin response? Usually the amplifier is tuned to be most sensitive to a certain rate of rise and fall of the response. If the change is either faster or slower, it will not be amplified as much. Thus, there are some areas of psychological research in which a person must be very sure of the response of the measuring instruments.

Problems of distortion also arise when psychological scaling is used. Suppose, as was described in an earlier section, we had scaled jokes from "hilarious" to "duds" by using the average ratings given by a group of judges. Can we be sure that the difference in funniness between a "2" joke and a "4" joke is the same as that between a "6" joke and an "8" joke? Probably not. Consequently, if we did an experiment to find how well jokes of different funniness were remembered and were testing some exact hypothesis (such as that there will be proportionally better remembering with proportional increase in funniness, we would not be too confident in deciding whether the shape of the curve supported the hypothesis or not. To do the experiment right, you would have to use a more sophisticated scaling technique than can be described in this book (see Torgerson, 1958). What you should know right now is that meaningful interpretation of shape of curves found by the use of behaviorally scaled variables always requires some proof that the variables are undistorted.

In the ideal experiment to test the hypothesis of an exact quantitative relation, there would be no distortion in measuring the independent and dependent variables. In any real measurement, there is always some distortion. If the distortion is so great that the relation found in the actual experiment might not represent the relation that would be found in the ideal experiment, internal validity has been weakened.

Earlier in this chapter, it was shown that in order to test any quantitative hypothesis—whether or not worded in actual quantitative terms—a sufficient number of levels of the independent variables must be used. Use of too few levels results in poorer representation of the *relation* between the independent and dependent variables. Internal validity was endangered not by unreliability or by confounding, but, rather, by incompleteness of the independent variable. Here it has been shown first that a group curve may misrepresent the individual and second that distorted measurements will show a false relation. In each case, the relation between the independent and dependent variables is *misread*. We now have three ways in which internal invalidity can come about, three reasons why the results of an actual experiment may fail to represent the relation between independent and dependent variables that would be found in an ideal experiment: (1) unreliability, (2) systematic confounding, and (3) misreading the relation.

SUMMARY

Three two-level experiments that might have been performed were shown to be spectacularly inadequate in comparison with an actual multilevel experiment. The advantages of the multilevel experiment were then spelled out.

First, it provides more effective controls in respect to internal validity than can the simpler experiments described in previous chapters. In some experiments using only two treatments, the independent variable logically is quantitative. Using only two levels of such a variable may result in missing the relation that would be found in an ideal quantitative experiment, in which an infinite number of levels are used. As more levels are used, there is a closer approach to this impossible experiment and, of course, improved internal validity. Also, there is better control over associative confounding. If an active level of the independent variable is compared with the zero (or inactive) level, it may indirectly introduce an active level of a secondary variable. Examples were being aware of a drug and making contact with an experimenter. On the other hand, a gradual change of the dependent variable with a gradual change of the independent variable is less likely to come about in this manner.

Multilevel experiments are superior to simpler experiments in a second way. Hypotheses may be tested which lead to greater understanding of the determination of behavior. First, in comparison with experiments using qualitative independent variables, there is better isolation of a unitary variable. A qualitative variable—such as reading vs. listening—can only be a "package." Opportunities for devising quantitative variables abound, including the use of behavioral scaling.

Further advances come through the testing of more ambitious hypotheses on the relation between the independent and dependent variables. An hypothesis that there will be a maximum (or minimum) value of the dependent variable for some intermediate level of the independent variable is often reasonable. It may follow from a theory of two underlying processes in which there is an aspect of opposition in relation to level of the independent variable. For example, it might be held that a negative process overtakes a positive one for very high levels of the independent variable. This analysis was applied to the imaginary multilevel experiment on the work ethic. The underlying variables were the feeling of "acting on the world" and the "noxious" effect. Another example is a theory holding that there are two positive underlying processes that are affected oppositely as level is increased. This analysis was applied to the experiment on interval between items in relation to memorizing of lists by college students and to the experiment relating strength of shock to the learning of a black-white discrimination by dancing mice. The underlying variables in the two experiments were stimulus differentiation and association.

Even more *detailed* experimental hypotheses are tested with multilevel independent variables. These are derived from models and theories of how step-by-step changes of the independent variable bring about changes in the dependent variable. Thus, from a scanning model of memory search, it was hypothesized that each equal absolute step of increase in size of memory set would bring about an equal absolute lengthening in search time: an absolute-absolute hypothesis. A proportional-absolute hypothesis was tested for the relation between number of alternatives and reaction time: For each equal proportional increase in number, there would be the same absolute increase in reaction time. This was based on a model for the most efficient way of making a decision. From a consideration of how sense organs translate physical stimulus energy to neural activity, a proportional-proportional hypothesis was predicted between amount of weight lifted and the subjective sensation of heaviness. In each case, the results supported the hypothesis: With appropriate scaling of axes, a straight line was found between the independent and dependent variables.

Experimental designs described previously may be used in multilevel experiments. The practical problem with between-groups designs is that too large a number of subjects may be needed. Within-subject controls are most useful when brief trials at different levels are given randomly within a large block of trials. When each trial is lengthy, as is usually the case when within-subject counterbalancing is used, the practical problem in a multilevel experiment is that too much time might be required for testing each subject.

These practical problems may be handled by using the across-subjects design of reverse counterbalancing. However, this particular design cannot control for *nonuniform transfer* effects from one trial to the next. Complete counterbalancing does provide such control but requires too many different sequences (and subject groups) usually to be practical. The typical method for across-subjects counterbalancing, which also controls for nonuniform transfer, is the Latin square, in which each level of the independent variable occurs once at each sequential position. Some further control is obtained by using only balanced squares, in which each level is preceded once by every other level. Still, any method of across-subjects counterbalancing cannot control for *range effects.* Low levels are more often preceded by high levels than by low levels, and high levels by low ones, introducing the threat of *asymmetrical transfer.* The other range effect described was that of *centering*, which occurs because only levels near the center of the range can be preceded by both lower and higher levels. In an experiment on block turning, this was found to favor levels near the center of the range of levels used. It was suggested that, when an across-subjects design is called for, a Latin square should be used instead of reverse counterbalancing, that sufficient rest be given between trials to avoid carryover of fatigue effects, and that overlapping ranges be used as a check for range effects.

Regardless of the experimental design used, there are two remaining threats to internal validity when testing exact hypotheses. One is that the shape of curve for the group might not represent that for any individual subject. In the ideal experiment, the same subject would be tested simultaneously at every level. The possibility of misrepresenting the relation that would be found is thus a source of internal invalidity. Between-groups designs are particularly vulnerable to this possibility. The danger can be minimized by matched subjects and homogeneous groups. Another way in which the relation between the independent and dependent variables might be misrepresented is through the distortion of the scales for measuring the independent and dependent variables. Again, the threat is to internal validity. This distortion is most apt to occur with two kinds of measurement. The first is when small physical quantities must be amplified. The second is when behavioral scaling is used.

In previous chapters, the threats to internal validity discussed were unreliability and systematic confounding. In this chapter, a new threat was identified, *misreading the relation* between the independent and dependent variables. This can come about through use of group curves that do not represent the individual, by testing at too few levels of the independent variable, and by making distorted measurements.

QUESTIONS

1. How does a multilevel experiment differ from those described in previous chapters?

2. What is meant by the statement that multilevel experiments provide controls for testing experimental hypotheses that could also be tested by two-treatment experiments?

3. Compare the understanding provided by an experiment with quantitative variation of the independent variable and one with only qualitatively different treatments.

4. What is meant by an experimental hypothesis of a maximum or minimum?

5. Why was the term *absolute-absolute* applied to Sternberg's experiment on memory search? What was the basis of his hypothesis?

6. Distinguish between the kind of quantitative experimental hypothesis tested by Hick (1952) on reaction time and by Harper and Stevens (1948) on subjective heaviness.

7. What are the practical reasons for using across-subjects counterbalancing rather than a between-groups design or within-subject counterbalancing?

8. What is meant by a Latin square design?

9. Is complete counterbalancing safe from nonuniform transfer effects? From range effects?

10. What are the threats to internal validity that remain using any design when testing the hypothesis of an exact relation between the independent and dependent variables?

11. The concept of the ideal experiment was reintroduced in respect to threats to internal validity other than unreliability or systematic confounding. How was this done? How would you define internal validity at this point?

STATISTICAL SUPPLEMENT

ONE-WAY ANALYSIS OF VARIANCE AND THE F-TEST

The t-test cannot be used to find the overall effect of the independent variable in a multilevel experiment. It can be used only to test the difference between two treatment means. A somewhat different approach and statistical test are required to find whether the means of the different levels generally differ from one another. The approach is called *analysis of variance*; statistical significance is evaluated by the F test. Since we are concerned with a single independent variable, the analysis is called a *one-way* analysis. In the statistical supplement to the next chapter, which examines experiments with two independent variables, the technique of two-way analysis of variance will be described.

Two Estimates of $\bar{\sigma}_X^2$

Let us consider an experiment, again on reaction time, in which four groups of subjects are used. The response is made to a tone, the independent variable is loudness (or more properly sound pressure). Four sound-pressure levels are used: 10 decibels (dB), 30 dB,

50 dB, and 70 dB. There are 17 subjects in each group, and the mean reaction time is found for each subject.

Suppose the null hypothesis were true. In an infinite experiment there would be the same value for $\overline{M}_1$, $\overline{M}_2$, $\overline{M}_3$, and $\overline{M}_4$, that is with an infinite number of subjects tested at each level. Still, of course, the mean reaction times for the different subjects given the same treatment would vary.

We are able to make two estimates of the parameter $\overline{\sigma}_x^2$ from the data of our experiment, again assuming the null hypothesis $\overline{M}_1 = \overline{M}_2 = \overline{M}_3 = \overline{M}_4$. One is based on the variation of reaction times among the subjects within each level. The *within-group* variation is simply pooled over the different levels. The other estimate is made from how much the individual group means vary from the *overall mean* of the experiment: $M_{1+2+3+4}$. Thus there is a *between-groups* estimate of $\overline{\sigma}_{\bar{x}}^2$ as well as a *within-group* estimate of $\overline{\sigma}_x^2$,

The Sampling Distribution of F

If the null hypothesis is true, there would be an identical estimate of $\overline{\sigma}_{\bar{x}}^2$—*in the long run*. Over the countless experiments, the average estimate from between-groups variation would be equal to that from within-group variation. In any particular experiment, including the one under consideration, we would not expect the estimates to be identical. In one experiment the two estimates might be very close; in another not so close.

When the two estimates are identical, their ratio equals 1:

$$\frac{\text{Between-groups estimate of } \overline{\sigma}_{\bar{X}}^2}{\text{Within-group estimate of } \overline{\sigma}_{\bar{X}}^2} = 1$$

This ratio is symbolized as F. In the expression above, the case is shown where $F = 1$. If the null hypothesis is wrong, the means for the different levels will differ much more than would be accounted for by unsystematic variation among subjects. The between-groups estimate will be larger than the within-group estimate; F will be *greater* than 1.

However, from experiment to experiment we would expect that the ratio F would vary from 1, even though on the average the value was 1 (as assumed by the null hypothesis). The distribution of values of F over countless experiments, assuming the null hypothesis to be true, is another *sampling distribution*. This sampling distribution can be drawn just as that for *t*. Here is an example:

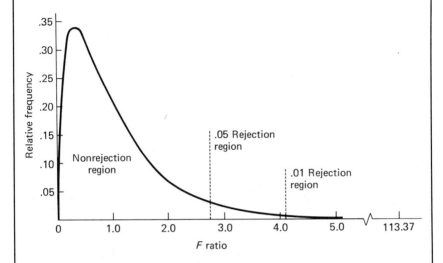

The question is whether, in a particular experiment, the value of *F* that is obtained is greater than that which exceeds the alpha level selected, usually .05 or .01. In other words, we will reject the null hypothesis only if it is sufficiently improbable that our obtained *F* would have occurred if the null were true. To do so, our *F* must certainly be larger than 1, much larger if there was a small number of subjects (or trials) or if there was much unsystematic variation.

Finding the Value of *F*

Let us now make a table that shows the terms needed to compute *F*.

	Level			
Term	1	2	3	4
M_x	M_1	M_2	M_3	M_4
ΣX^2	ΣX_1^2	ΣX_2^2	ΣX_3^2	ΣX_4^2
n	n_1	n_2	n_3	n_4

Since we have already made some computations on four sets of data, let us pretend that they were gathered for the experiment on levels of loudness on reaction. Call Treatment C, Level 1; call Treatment D, Level 2; call Treatment A, Level 3; call Treatment

B, Level 4. (This will save us a lot of computing. Moreover it will make mean reaction time shorter as loudness increases, as it should!) We thus have already computed the important terms (see Chapter 6).

		Sound Level		
Term	1	2	3	4
M_x	265	250	185	162
Σx^2	4673	5391	5808	4306
n	17	17	17	17

Sum of Squares, Within Groups   The sum of squares, (SS), within groups, (WG), will be used in finding the within-groups estimate of $\bar{\sigma}_x^2$. It is found by simply adding across the Σ_x^2 row, thus:

$$SS_{WG} = \Sigma x_1^2 + \Sigma x_2^2 + \Sigma x_3^2 \qquad \text{(Formula 7.1)}$$

Here it is

$$SS_{WG} = 4673 + 5391 + 5808 + 4306 = 20{,}178$$

Sum of Squares, Between Groups   The sum of squares, between groups, will be used in finding the between-groups estimate of $\bar{\sigma}_x^2$. To find it, you first compute the overall (total, "tot") mean for the four treatments:

$$M_{\text{tot}} = \frac{M_1 + M_2 + M_3 \cdots}{k} \qquad \text{(Formula 7.2)}$$

where k is the number of groups. Here

$$M_{\text{tot}} = \frac{265 + 250 + 185 + 162}{4} = \frac{862}{4} = 215.5$$

Next, the difference is found between each of the separate means and the overall mean. These differences are symbolized as d. Thus

$$d_1 = M_1 - M_{\text{tot}}, \qquad d_2 = M_2 - M_{\text{tot}} \cdots \qquad \text{(Formula 7.3)}$$

Here

$$d_1 = 265 - 215.5 = +49.5, \qquad d_2 = 250 - 215.5 = +34.5$$

$$d_3 = 185 - 215.5 = -30.5, \qquad d_4 = 162 - 215.5 = -53.5$$

The sum of squares, between groups, is simply the sum of the squared d values times the number of cases (n) per treatment.

$$SS_{BG} = n(d_1^2 + d_2^2 + d_3^2 \cdots) \qquad \text{(Formula 7.4)}$$

Here

$$SS_{BG} = 17(2450.25 + 1190.25 + 930.25 + 2862.25)$$

$$= 17(7433)$$

$$= 126,361$$

Mean Square, Within Groups.   The estimate of $\bar{\sigma}_x^2$ based on within-groups variation is called the *mean square, within groups*. It is found by dividing the sum of squares within groups by the combined number of degrees of freedom for the means of the several groups. Thus, it is $(n_1 - 1) + (n_2 - 1) + (n_3 - 1), \ldots$.

Since there are k treatments and N subjects in all,

$$df_{WG} = N - k \qquad \text{(Formula 7.5)}$$

Here

$$df_{WG} = 68 - 4 = 64$$

As stated,

$$MS_{WG} = \frac{SS_{WG}}{N - k} \qquad \text{(Formula 7.6)}$$

Here

$$MS_{WG} = \frac{20,178}{64} = 315$$

Mean Square, Between Groups.   The estimate of $\bar{\sigma}_{\bar{x}}^2$ based on between-groups variation is called *mean square, between groups*. It is found by dividing the sum of squares between groups by the number of degrees of freedom for the overall mean when computed from the means for the different treatments.

$$df_{BG} = k - 1 \qquad \text{(Formula 7.7)}$$

Here

$$df_{BG} = 4 - 1 = 3$$

As stated,

(Formula 7.8)

$$MS_{BG} = \frac{SS_{BG}}{df_{BG}}$$

Here

$$MS_{BG} = \frac{126,361}{3} = 42,120$$

The F-Ratio.   The final step in computing F is to divide the mean square, between groups, by the mean square, within groups. Remember, the larger this ratio the more probable it will be that the null hypothesis may be rejected.

$$F = \frac{MS_{BG}}{MS_{WG}}$$

(Formula 7.9)

Here

$$F = \frac{42,120}{315}$$

$$= 133.71$$

Rejecting or Not Rejecting the Null Hypothesis

Our obtained value of F is seen to be far off to the right in the drawing of the F distribution near the beginning of this statistical supplement. Obviously, this large an F-ratio would be extremely rare if the null hypothesis were true—that in countless experiments the ratio would average out at 1. We can make sure that we want to reject the null hypothesis by finding the critical value in Statistical Table 3 at the end of this supplement.

Since the distribution will have a different shape according to the number of degrees of freedom in the numerator and the denominator, it is set up with several vertical columns and many horizontal rows. Each *column* contains the critical values of F, for alpha levels .05 and .01 for a certain number of degrees of freedom in the *nu-*

merator of the F-ratio. Each row does the same for degrees of freedom in the denominator.

To use Statistical Table 3 for our F of 133.71 with 3 *df* in the numerator and 64 *df* in the denominator, we go to Column 3 and Row 65 (which is close enough). The entry 2.75 tells us the value of F required to reject the null hypothesis at the .05 level; the entry 4.10 tells us the value of F required to reject the null hypothesis at the .01 level. Lines are drawn at these values in the drawing of the F distribution. The rejection region for each of these alpha levels lies to the right of the line. Of course there is no need to draw the distribution as we only have to use the table of critical values. Here we would state that $p < .01$.

Analysis of Variance Table

The method that has just been described is called *analysis of variance* (or ANOVA, in computerese). All of the variance in the data, in fact, has been analyzed into portions. You could have subtracted the overall mean from each subject's score and squared each of the 68 differences. Adding them all together gives the sum of squares, total (SS_{tot}). Now, if you add together the sum of squares, within groups, and the sum of squares, between groups— and have not made any mistakes—this sum will equal the sum of squares, total.

In presenting the results of an analysis of variance, it is customary to use a table of the various sums of squares and mean squares. Here is how our results might be shown:

Analysis of Variance,
Experiment on Loudness of Stimuli and Reaction Time

Source of Variance	SS	df	MS	F	p
Between Loudness Levels	126,361	3	42,120	133.71	< .01
Within Loudness Levels	20,178	64	315		
Total	146,539	67			

PROBLEM: Do an analysis of variance from the following data relating number of problems solved to size of monetary reward. Complete the analysis of variance table. The data are from separate groups of subjects.

Reward (from Low to High)

Level 1	Level 2	Level 3	Level 4	Level 5	Level 6
10	8	12	12	24	19
11	10	17	15	16	18
9	16	14	16	22	27
13	13	9	16	18	25
7	12	16	19	20	24

Answer:

Source of Variance	SS	df	MS	F	p
Between Loudness Levels	590.8	5	118.16	12.64	$< .01$
Within Loudness Levels	224.4	24	9.35		
Total	815	29			

STATISTICAL TABLE 3
Critical Values of F for
Rejecting the Null Hypothesis
(.05 lightface, **.01 boldface**)[a]

Degrees of freedom for denominator	Degrees of Freedom for Numerator									
	1	2	3	4	5	6	7	8	9	10
1	161	200	216	225	230	234	237	239	241	242
	4052	**4999**	**5403**	**5625**	**5764**	**5859**	**5928**	**5981**	**6022**	**6056**
2	18.51	19.00	19.16	19.25	19.30	19.33	19.36	19.37	19.38	19.39
	98.49	**99.01**	**99.17**	**99.25**	**99.30**	**99.33**	**99.34**	**99.36**	**99.38**	**99.40**
3	10.13	9.55	9.28	9.12	9.01	8.94	8.88	8.84	8.81	8.78
	34.12	**30.81**	**29.46**	**28.71**	**28.24**	**27.91**	**27.67**	**27.49**	**27.34**	**27.23**
4	7.71	6.94	6.59	6.39	6.26	6.16	6.09	6.04	6.00	5.96
	21.20	**18.00**	**16.69**	**15.98**	**15.52**	**15.21**	**14.98**	**14.80**	**14.66**	**14.54**
5	6.61	5.79	5.41	5.19	5.05	4.95	4.88	4.82	4.78	4.74
	16.26	**13.27**	**12.06**	**11.39**	**10.97**	**10.67**	**10.45**	**10.27**	**10.15**	**10.05**
6	5.99	5.14	4.76	4.53	4.39	4.28	4.21	4.15	4.10	4.06
	13.74	**10.92**	**9.78**	**9.15**	**8.75**	**8.47**	**8.26**	**8.10**	**7.98**	**7.87**
7	5.59	4.74	4.35	4.12	3.97	3.87	3.79	3.73	3.68	3.63
	12.25	**9.55**	**8.45**	**7.85**	**7.46**	**7.19**	**7.00**	**6.84**	**6.71**	**6.62**
8	5.32	4.46	4.07	3.84	3.69	3.58	3.50	3.44	3.39	3.34
	11.26	**8.65**	**7.59**	**7.01**	**6.63**	**6.37**	**6.19**	**6.03**	**5.91**	**5.82**
9	5.12	4.26	3.86	3.63	3.48	3.37	3.29	3.23	3.18	3.13
	10.56	**8.02**	**6.99**	**6.42**	**6.06**	**5.80**	**5.62**	**5.47**	**5.35**	**5.26**
10	4.96	4.10	3.71	3.48	3.33	3.22	3.14	3.07	3.02	2.97
	10.04	**7.56**	**6.55**	**5.99**	**5.64**	**5.39**	**5.21**	**5.06**	**4.95**	**4.85**
11	4.84	3.98	3.59	3.36	3.20	3.09	3.01	2.95	2.90	2.86
	9.65	**7.20**	**6.22**	**5.67**	**5.32**	**5.07**	**4.88**	**4.74**	**4.63**	**4.54**
12	4.75	3.88	3.49	3.26	3.11	3.00	2.92	2.85	2.80	2.76
	9.33	**6.93**	**5.95**	**5.41**	**5.06**	**4.82**	**4.65**	**4.50**	**4.39**	**4.30**
13	4.67	3.80	3.41	3.18	3.02	2.92	2.84	2.77	2.72	2.67
	9.07	**6.70**	**5.74**	**5.20**	**4.86**	**4.62**	**4.44**	**4.30**	**4.19**	**4.10**
14	4.60	3.74	3.34	3.11	2.96	2.85	2.77	2.70	2.65	2.60
	8.86	**6.51**	**5.56**	**5.03**	**4.69**	**4.46**	**4.28**	**4.14**	**4.03**	**3.94**
15	4.54	3.68	3.29	3.06	2.90	2.79	2.70	2.64	2.59	2.55
	8.68	**6.36**	**5.42**	**4.89**	**4.56**	**4.32**	**4.14**	**4.00**	**3.89**	**3.80**
16	4.49	3.63	3.24	3.01	2.85	2.74	2.66	2.59	2.54	2.49
	8.53	**6.23**	**5.29**	**4.77**	**4.44**	**4.20**	**4.03**	**3.89**	**3.78**	**3.69**
17	4.45	3.59	3.20	2.96	2.81	2.70	2.62	2.55	2.50	2.45
	8.40	**6.11**	**5.18**	**4.67**	**4.34**	**4.10**	**3.93**	**3.79**	**3.68**	**3.59**
18	4.41	3.55	3.16	2.93	2.77	2.66	2.58	2.51	2.46	2.41
	8.28	**6.01**	**5.09**	**4.58**	**4.25**	**4.01**	**3.85**	**3.71**	**3.60**	**3.51**
19	4.38	3.52	3.13	2.90	2.74	2.63	2.55	2.48	2.43	2.38
	8.18	**5.93**	**5.01**	**4.50**	**4.17**	**3.94**	**3.77**	**3.63**	**3.52**	**3.43**
20	4.35	3.49	3.10	2.87	2.71	2.60	2.52	2.45	2.40	2.35
	8.10	**5.85**	**4.94**	**4.43**	**4.10**	**3.87**	**3.71**	**3.56**	**3.45**	**3.37**
21	4.32	3.47	3.07	2.84	2.68	2.57	2.49	2.42	2.37	2.32
	8.02	**5.78**	**4.87**	**4.37**	**4.04**	**3.81**	**3.65**	**3.51**	**3.40**	**3.31**
22	4.30	3.44	3.05	2.82	2.66	2.55	2.47	2.40	2.35	2.30
	7.94	**5.72**	**4.82**	**4.31**	**3.99**	**3.76**	**3.59**	**3.45**	**3.35**	**3.26**
23	4.28	3.42	3.03	2.80	2.64	2.53	2.45	2.38	2.32	2.28
	7.88	**5.66**	**4.76**	**4.26**	**3.94**	**3.71**	**3.54**	**3.41**	**3.30**	**3.21**
24	4.26	3.40	3.01	2.78	2.62	2.51	2.43	2.36	2.30	2.26
	7.82	**5.61**	**4.72**	**4.22**	**3.90**	**3.67**	**3.50**	**3.36**	**3.25**	**3.17**
25	4.24	3.38	2.99	2.76	2.60	2.49	2.41	2.34	2.28	2.24
	7.77	**5.57**	**4.68**	**4.18**	**3.86**	**3.63**	**3.46**	**3.32**	**3.21**	**3.13**
26	4.22	3.37	2.98	2.74	2.59	2.47	2.39	2.32	2.27	2.22
	7.72	**5.53**	**4.64**	**4.14**	**3.82**	**3.59**	**3.42**	**3.29**	**3.17**	**3.09**

Degrees of freedom for denominator	Degrees of Freedom for Numerator									
	1	2	3	4	5	6	7	8	9	10
27	4.21	3.35	2.96	2.73	2.57	2.46	2.37	2.30	2.25	2.20
	7.68	**5.49**	**4.60**	**4.11**	**3.79**	**3.56**	**3.39**	**3.26**	**3.14**	**3.06**
28	4.20	3.34	2.95	2.71	2.56	2.44	2.36	2.29	2.24	2.19
	7.64	**5.45**	**4.57**	**4.07**	**3.76**	**3.53**	**3.36**	**3.23**	**3.11**	**3.03**
29	4.18	3.33	2.93	2.70	2.54	2.43	2.35	2.28	2.22	2.18
	7.60	**5.42**	**4.54**	**4.04**	**3.73**	**3.50**	**3.33**	**3.20**	**3.08**	**3.00**
30	4.17	3.32	2.92	2.69	2.53	2.42	2.34	2.27	2.21	2.16
	7.56	**5.39**	**4.51**	**4.02**	**3.70**	**3.47**	**3.30**	**3.17**	**3.06**	**2.98**
32	4.15	3.30	2.90	2.67	2.51	2.40	2.32	2.25	2.19	2.14
	7.50	**5.34**	**4.46**	**3.97**	**3.66**	**3.42**	**3.25**	**3.12**	**3.01**	**2.94**
34	4.13	3.28	2.88	2.65	2.49	2.38	2.30	2.23	2.17	2.12
	7.44	**5.29**	**4.42**	**3.93**	**3.61**	**3.38**	**3.21**	**3.08**	**2.97**	**2.89**
36	4.11	3.26	2.86	2.63	2.48	2.36	2.28	2.21	2.15	2.10
	7.39	**5.25**	**4.38**	**3.89**	**3.58**	**3.35**	**3.18**	**3.04**	**2.94**	**2.86**
38	4.10	3.25	2.85	2.62	2.46	2.35	2.26	2.19	2.14	2.09
	7.35	**5.21**	**4.34**	**3.86**	**3.54**	**3.32**	**3.15**	**3.02**	**2.91**	**2.82**
40	4.08	3.23	2.84	2.61	2.45	2.34	2.25	2.18	2.12	2.07
	7.31	**5.18**	**4.31**	**3.83**	**3.51**	**3.29**	**3.12**	**2.99**	**2.88**	**2.80**
42	4.07	3.22	2.83	2.59	2.44	2.32	2.24	2.17	2.11	2.06
	7.27	**5.15**	**4.29**	**3.80**	**3.49**	**3.26**	**3.10**	**2.96**	**2.86**	**2.77**
44	4.06	3.21	2.82	2.58	2.43	2.31	2.23	2.16	2.10	2.05
	7.24	**5.12**	**4.26**	**3.78**	**3.46**	**3.24**	**3.07**	**2.94**	**2.84**	**2.75**
46	4.05	3.20	2.81	2.57	2.42	2.30	2.22	2.14	2.09	2.04
	7.21	**5.10**	**4.24**	**3.76**	**3.44**	**3.22**	**3.05**	**2.92**	**2.82**	**2.73**
48	4.04	3.19	2.80	2.56	2.41	2.30	2.21	2.14	2.08	2.03
	7.19	**5.08**	**4.22**	**3.74**	**3.42**	**3.20**	**3.04**	**2.90**	**2.80**	**2.71**
50	4.03	3.18	2.79	2.56	2.40	2.29	2.20	2.13	2.07	2.02
	7.17	**5.06**	**4.20**	**3.72**	**3.41**	**3.18**	**3.02**	**2.88**	**2.78**	**2.70**
55	4.02	3.17	2.78	2.54	2.38	2.27	2.18	2.11	2.05	2.00
	7.12	**5.01**	**4.16**	**3.68**	**3.37**	**3.15**	**2.98**	**2.85**	**2.75**	**2.66**
60	4.00	3.15	2.76	2.52	2.37	2.25	2.17	2.10	2.04	1.99
	7.08	**4.98**	**4.13**	**3.65**	**3.34**	**3.12**	**2.95**	**2.82**	**2.72**	**2.63**
65	3.99	3.14	2.75	2.51	2.36	2.24	2.15	2.08	2.02	1.98
	7.04	**4.95**	**4.10**	**3.62**	**3.31**	**3.09**	**2.93**	**2.79**	**2.70**	**2.61**
70	3.98	3.13	2.74	2.50	2.35	2.23	2.14	2.07	2.01	1.97
	7.01	**4.92**	**4.08**	**3.60**	**3.29**	**3.07**	**2.91**	**2.77**	**2.67**	**2.59**
80	3.96	3.11	2.72	2.48	2.33	2.21	2.12	2.05	1.99	1.95
	6.96	**4.88**	**4.04**	**3.56**	**3.25**	**3.04**	**2.87**	**2.74**	**2.64**	**2.55**
100	3.94	3.09	2.70	2.46	2.30	2.19	2.10	2.03	1.97	1.92
	6.90	**4.82**	**3.98**	**3.51**	**3.20**	**2.99**	**2.82**	**2.69**	**2.59**	**2.51**
125	3.92	3.07	2.68	2.44	2.29	2.17	2.08	2.01	1.95	1.90
	6.84	**4.78**	**3.94**	**3.47**	**3.17**	**2.95**	**2.79**	**2.65**	**2.56**	**2.47**
150	3.91	3.06	2.67	2.43	2.27	2.16	2.07	2.00	1.94	1.89
	6.81	**4.75**	**3.91**	**3.44**	**3.14**	**2.92**	**2.76**	**2.62**	**2.53**	**2.44**
200	3.89	3.04	2.65	2.41	2.26	2.14	2.05	1.98	1.92	1.87
	6.76	**4.71**	**3.88**	**3.41**	**3.11**	**2.90**	**2.73**	**2.60**	**2.50**	**2.41**
400	3.86	3.02	2.62	2.39	2.23	2.12	2.03	1.96	1.90	1.85
	6.70	**4.66**	**3.83**	**3.36**	**3.06**	**2.85**	**2.69**	**2.55**	**2.46**	**2.37**
1000	3.85	3.00	2.61	2.38	2.22	2.10	2.02	1.95	1.89	1.84
	6.66	**4.62**	**3.80**	**3.34**	**3.04**	**2.82**	**2.66**	**2.53**	**2.43**	**2.34**
∞	3.84	2.99	2.60	2.37	2.21	2.09	2.01	1.94	1.88	1.83
	6.64	**4.60**	**3.78**	**3.32**	**3.02**	**2.80**	**2.64**	**2.51**	**2.41**	**2.32**

FACTORIAL
EXPERIMENTS

In the study by David Gaffan (1974) in which he transected the fornix of six rhesus monkeys and performed a control operation on six others (described in Chapter 5), a test was being made of a precisely stated experimental hypothesis: Damage to the hippocampal system of the brain impairs *recognition memory*. If this were shown to be true, a considerable advance would be made in understanding the amnesia that occurs in many cases of accidental brain damage to human beings.

Notice how specific the hypothesis is. The effect is on *memory*—not on anything else. The effect is on *recognition* memory—not on any other kind. In order to isolate the effect, an experiment would clearly need some well thought-out controls. Let us now see how all this was done.

At first, nothing was done. The monkeys were allowed to recover from their operation for two weeks. After that, they were taught to play the laboratory version of the old-time carnival shell-and-pea game (but with no sleight-of-hand tricks). It is called *matching to sample*. Here is how it is played. A tray containing three food wells, side by side, is put before a monkey. On a *sample* trial, the middle well is covered by a small object, perhaps a little wooden boat. All that the monkey has to do is to lift the object so that it can scoop out the Sugar Puffs (a dry cereal made by Quaker Oats) hidden underneath in the well. Ten seconds later comes the *matching* trial. The monkey is presented with the tray with two objects covering the two side wells; there is nothing in the middle well now. One of the side wells is covered by the little boat (if that was the object used on the sample trial), and the other side well has a different small object covering it—perhaps a toy telephone. If the monkey picks up the little boat, it finds some more Sugar Puffs and has, of course, made a correct choice. If it picks up the toy telephone, it finds nothing in the food well—an error.

For each monkey, this game was played until 81 correct choices were made out of 90 consecutive matching to sample trials. This comprised a 3-day period. Achievement of this level of accuracy satisfied the experimenter's *criterion* of effective performance. The experimenter had at his disposal a bank of 300 "junk" items, such as toys, pans, electric relays, etc. They were all quite different from each other, and there were so many that the experimenter could go for 5 days without a repetition. Needless to say, the left and right food wells were made correct randomly on successive trials. Both groups of monkeys learned to match from sample about equally quickly. Criterion was reached by all monkeys in between 330 and 600 paired trials. We should note that the monkeys received their normal diet all through the experiment; the Sugar Puffs were an added "treat."

The two groups were then tested on several tasks, so the investigation was divided into subexperiments. One of the tasks was called *delays*. In the initial training, only the 10-sec delay between sample and matching trials was used. Now there were three delays: 10, 70, and 130 sec. That is, for the longest delay, an animal had to remember the object that covered the food reward for more than 2 min. The results are shown in Figure 8.1.

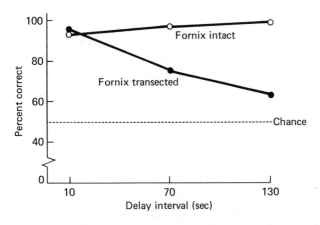

Figure 8.1 Performance on the delay task by fornix-transected and fornix-intact (operated control) monkeys.

It can be seen that there was really no difference in performance between the fornix-transected animals and operated controls for the 10-sec delay. This, of course, is no surprise. It is just like the practice task, which the two groups learned about equally well. However, as the delay became longer, the groups differed more. Whereas the control animals still averaged more than 90 percent correct responses at the 130-sec delay, the fornix-transected group dropped to 65 percent correct. This is not much better than the 50 percent correct they would have achieved by chance. The good performance at the 10-sec delay confirms the adequacy of *registration* by the fornix-transected animals. That is to say, the correct object was known to them immediately after the sample trial. However, the control group continued to remember the correct object while the experimental group forgot it. Thus it was shown that the deficit of the fornix-transected animals was not in registration but was in retention. If there had been testing only at the 130-sec interval, it would have been impossible to rule out registration. A second independent variable, in addition to the independent variable of *state of fornix*, was necessary in order to control for the possibility that the deficit was in registration rather than memory. The second independent variable was *delay interval*. There remained the problem of showing that the deficit was in *recognition* memory. How it was solved will be described in the next section.

In the previous chapters, each experiment used one independent variable and one dependent variable. In the subexperiment by Gaffan, there was more than one independent variable related to the dependent variable. In addition to state of fornix, there was delay interval. Without this variable—using only a long retention interval—there would be no way of knowing that the deficit was really in memory rather than in registration. The use of a second independent variable is seen to function as a special kind of control to isolate the effect of the independent variable of interest. That is one reason that experiments are devised with more than one independent variable.

We shall see that the other main use is in the testing of more complicated experimental hypotheses than we have encountered previously, namely those concerning the way in which independent variables work in combination to affect behavior. We will call them *combinational* hypotheses. The term *factorial experiment* is used when there is more than one independent variable. All that means is that each independent variable is a (possible) factor in behavior. Since, as we know, behavior is complexly determined, the testing of combinational hypotheses represents another advance in our attempts at understanding.

You might be wondering whether there are experiments with more than one *dependent* variable so that a relation may be found between the independent variable and each of the dependent variables. The answer is yes. There are also experiments with both more than one independent and more than one dependent variable. However, these experiments are fairly new in psychology and are somewhat out of our depth from a statistical point of view. The term used for these advanced experiments is *multivariate*. There are some of you who will go on to master these methods. At the present time, you will find that the large majority of experiments in the literature do not go beyond the factorial.

The term *factorial design* is used to describe how these experiments are set up. With this type of design, we have a new dimension for classifying experimental designs. Near the end of this chapter, the scheme by which experimental designs may be classified will be presented so that you will be able to keep the various possibilities in mind.

You will find, as you read experimental articles being written at the present time, that almost every one uses factorial design. Some knowledge of what is being done and what the terms mean is essential if you are to comprehend these articles at all. After you finish this chapter, you will understand what is meant by *main effects* and *interactions*. Perhaps you will run across studies that could have been strengthened by use of a second independent variable. There is no reason at all for your being unable to go on to this kind of design in your own experiments. However, you may have to stretch your mind just a little to think of two independent variables at the same time. You will not be asked in this book to master experiments with more than two independent variables. Fortunately, few investigators try to get beyond three, which will be within your reach if you continue to involve yourself in the experimental part of psychology.

Here are the main areas in which you may expect to be questioned at the end of the chapter:

1. Basic terms: *factorial experiment, main effects, interaction.*
2. Factorial experiments as controls when testing single-relation hypotheses.
3. The testing of combinational hypotheses: expected interactions.
4. Classification of experimental designs.

MAIN EFFECTS AND INTERACTIONS

We will start this section by describing the other subexperiment by Gaffan (actually, combined from two of his "tasks") to be considered in this chapter. For the present, our description will be for the purpose of showing concretely what is meant by *main effects* and by *interactions*. In the section that follows, there will be a discussion of the way in which these subexperiments provided effective experimental controls.

An Experiment To Separate Recognition Memory and Association Memory

Two tasks were used in the subexperiment to show that it was recognition memory—not associative memory—that was impaired by transection of the fornix. One of the tasks was called *association*. In this task, each set of acquisition trials was followed by a retention test. Ten objects were used on the 10 acquisition trials of a set, which were given one after another. On five of the trials, the object covered the Sugar Puffs; on the other five trials there was just an empty food well under the object. The retention test followed immediately. Each of the 10 objects was placed over the right-end food well at 20-sec intervals, exactly in the order originally presented. The left-end food well was covered with a brass disk, never used on the acquisition trials. If the object covering the right-end food well was one of the five that had covered the reward on an acquisition trial, it again covered the Sugar Puffs. However, if it was one of those which covered an empty well previously, the food was located under the brass disk. Thus, the correct response was to lift an object that had previously covered food if that was present, but to lift the brass disk otherwise. Three sets of trials were run each day on an animal. On the fifth and last day the two groups performed almost identically, making slightly more than 80 percent correct responses.

The second of the tasks in this subexperiment was called *yes/no recognition*. It was much the same as association but had a critical difference. As before, five rewarded items were used on the acquisition trials, but now there were no nonrewarded ones. Instead, five new items were the nonrewarded ones on the retention test that followed. Again, it was correct to pick up an object covering the right-end food well if the object had done so on the acquisition trials. If the object was one never seen before, the brass disk at the left end covered the food. Now, on the last test day, there was a great difference between the two groups of monkeys. The control group did at least as well as on the association task. However, the fornix-transected group made barely more than 60 percent correct responses, far down from the previous 80 percent.

In the association task, the requirement for a correct response was to remember whether each object in the test was associated with reward or nonreward during the acquisition trials. In the recognition task, more was needed, since the wrong choices were all new objects. Thus the additional requirement was to recognize an object as familiar or unfamiliar. The fornix-transected group was found to have perfectly good associative memory but to fail at recognition.

Measuring Main Effects

In the factorial experiment just described, we can find the separate effects of each of the two independent variables. Since each independent variable had only two levels or treatments, this is very simple. It is the overall difference between the means for the two levels. For the control animals, with the intact fornix, there was a mean of 83 percent of correct responses on the associative task, in which there was previous exposure of negative objects and a mean of 88 percent on the recognition task, in which there was no previous exposure. The overall mean for intact fornix was thus 85.5 percent. For the monkeys with a transected fornix, the means with previous exposure were 82 percent and 62 percent, for an overall mean 72.0 percent. The main effect for the independent variable, state of fornix, was the difference between these two means, 13.5 percent. We shall continue this analysis as though there were always only two levels of each independent variable so that it will be easy to follow. The same logic holds for multilevel experiments; however, the computations are more involved.

Instead of writing all this out it may be shown more directly in a table or in a graph. In Table 8.1, the breakdown of values is shown by the four inside numbers: 82, 62, 83, and 88. It is typical to omit the unit of measurement—here, percent correct responses. The two overall means for the transected and intact fornix groups are shown at the bottom margin. The main effect, state of fornix, is the difference between these two *marginal* means, 85.5 and 72.0, which equals 13.5.

TABLE 8.1
**Computation of the main effects,
state of fornix and previous exposure
to negative objects, and of their interaction**

Previous Exposure to Negative Objects	State of Fornix Transected	Intact	Mean
Yes	82	83	82.5
No	62	88	75.0
Mean	72.0	85.5	76.25

Computations

Main effect: State of fornix

$\text{Mean}_{\text{intact}} - \text{Mean}_{\text{transected}}$
$85.5 - 72.0 = 13.5$

Main effect: Previous exposure to negative objects

$\text{Mean}_{\text{yes}} - \text{Mean}_{\text{no}}$
$82.5 - 75.0 = 7.5$

Interaction: Fornix × Previous exposure

(No previous exposure, intact − transected) − (Yes previous exposure, intact − transected)

$$(88 - 62) \qquad\qquad - (83 - 82)$$
$$26 \qquad\qquad\quad - 1 = 25$$

or

(Transected, yes −no) − (Intact, yes −no)

$$(82 - 62) \qquad\qquad - (83 - 88)$$
$$20 \qquad\qquad\quad - (-5) = 25$$

In the same way, the two means on the right-hand margin are the overall values for the "yes" level of *Previous Exposure to Negative Objects* and the "no" level. This main effect, is seen to be a good deal smaller: 82.5 minus 75.0 equals only 7.5. You see the formulas for computing the two main effects directly below the table itself.

The main effects are also represented in the upper graph of Figure 8.2. In addition to the solid lines for the fornix-intact group and the fornix-transected group, there are two pairs of dashed lines extended to the right. One line comes from the middle of the fornix-intact line and so is at the overall mean value of that group, 85.5. It is paired with the line from the middle of the fornix-transected group, and—of course—is at that group's overall mean value, 72.0. Far on the right-hand side, the gap or distance between these two lines is seen to be 13.5, just as determined from the table. One line of the second pair comes from the midpoint of the "yes" previous exposure task and so is at its overall mean value, 82.5. Its paired line comes from the midpoint of "no" previous exposure, 75.0. The comparison is made just to the right, and the distance is 7.5, again as determined from the table. Thus, each main effect may be represented simply as the distance between two lines.

Measuring Interactions

Now we come to the way in which an interaction is measured. It is not a simple difference. Rather, it is the difference between two differences. To compute it, we use the values inside the table, not the marginal ones. Let us ask the question all over again of whether animals with an intact fornix performed better than those with a transection. We find that it depends on which task we are talking about. With no previous exposure, the intact group did much better; with previous exposure, the two groups did equally well. Look again at the inside values in Table 8.1 to see how a quantitative measure of this interaction is obtained. On the task with no previous exposure, for the intact versus the transected group there were 88 as compared with 62 percent correct responses—a difference of 26 percent. On the task with previous exposure, there were 83 as compared to 82 percent correct responses—a difference of 1 percent. The interaction is the difference between these two differences, or 26 minus 1. The computation is shown in the formula directly under the main effects. The full name for the interaction in this case is "state of fornix *times* previous exposure to negative objects." Usually some shorter name is used: here, "fornix × previous exposure."

What the interaction has just told us is the extent to which the effect of the state of the fornix depends on the task. This may be turned around by asking again whether performance is better with or without previous exposure to negative objects. Now, the answer is going to depend on whether we are talking about animals with a transected fornix or ones with an intact fornix. With a transected fornix, previous exposure was helpful: 82 vs. 62 percent correct responses, a difference of 20 percent. For the intact animals, not only was the difference reduced, but it even went slightly the other way, 83 vs. 88 percent. If we do our subtraction in the same direction as for the transected animals, as is shown in the formula at the botton of Table 8.1, we see that we must subtract −5 from 20. Again, the interaction comes out to be 25. Either way the question is asked, the difference on one variable *depends* on the level of the other variable. That is really what is meant by interaction between two independent variables.

The two ways of computing the interaction are illustrated in the bottom graph of Figure 8.2. The solid lines are just as drawn in the upper graph. Two pairs of dashed lines are compared on the left. The distance between the intact and transected groups is shown to be 26 on the "no" previous experience task but only 1 on the "yes" task. The difference between these two distances is 25. Over on the right, there is a lower pair of dotted lines that shows a difference distance of 20 for the transected group on the two tasks. The difference for the intact group is the other way around, better performance on the "no" task, but with a value of only 5. The difference between the two distances is again 25, since to subtract a negative number means to add the corresponding positive number.

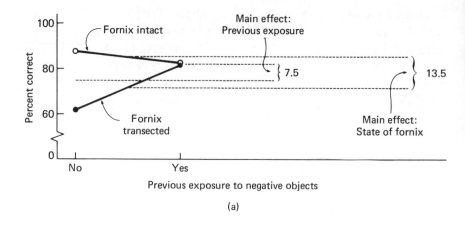

(a)

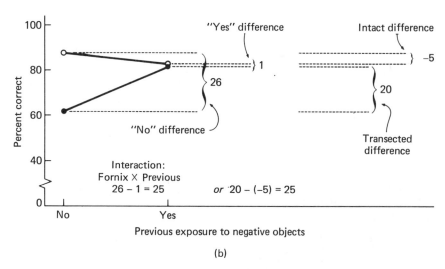

(b)

Figure 8.2 Graphic measurement of the main effects, state of fornix and previous exposure to negative objects, and their interaction.

Kinds of Interaction

In the subexperiment by Gaffan that has been under discussion, we may say that a moderate main effect was found for the independent variable state of fornix (13.5), a much smaller main effect for the independent variable previous exposure to negative objects (7.5), and a very strong interaction between state of fornix and previous exposure (25). Let us see what we would say if the experiment had turned out in different ways. In Figure 8.3 are shown a number of possible outcomes, given both in tabular form and as graphs. In (a) the main effects are a little larger than found

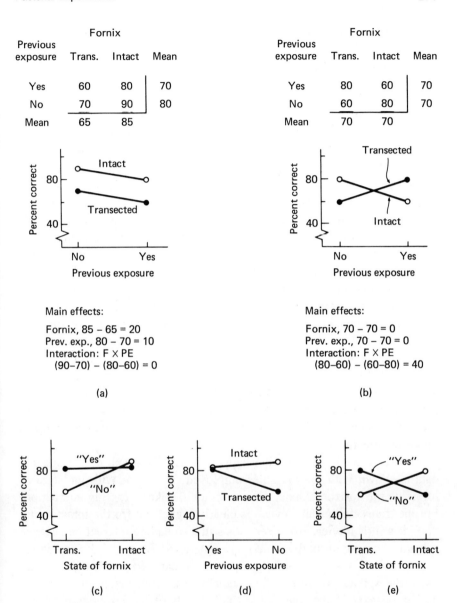

Figure 8.3 Possible outcomes of the experiment with two independent variables, state of fornix and previous exposure to negative effects: main effects and interactions.

in the real study: 20 and 10, for state of fornix and previous exposure respectively. However, there is a zero value of interaction. (Note the "shorthand" representation, $F \times PE$.) When this occurs, it is seen that the two lines are parallel, unlike that for the actual data represented in Figure 8.2. We would say that (a) represents *no* interaction between the independent variables.

In contrast with the *zero* interaction portrayed in (a), there is even more interaction shown in the make-believe outcome presented in (b) than there was in the actual experiment. Here there is a *crossing* interaction. The actual data showed a *spreading* interaction. With the *crossing* interaction, there is just the opposite difference between the fornix-transected and intact animals with and without previous exposure. Whereas the transected group did better than the intact group with previous exposure, they did worse without it. This also may be stated: There was better performance with previous exposure than without for the transected group, but for the intact group there was better performance without previous exposure than with it. The data of (b) are so represented in (e) that this comparison may be made. All that has been done is to put the state of fornix variable on the horizontal axis and to use two lines for the two levels of the previous exposure variable—"yes" or "no." For a crossing interaction, much the same picture will be seen whichever of the two ways the graph is drawn. You will notice in the computations at the bottom of (b) that each of the main effects has a value of zero. Overall, there is no difference between the transected and intact animals; there is no difference between the task with previous experience and that without it. However, a very large value is found for interaction, 40. That is because the value of -20 was subtracted from 20, the changes being in opposite directions.

In (c) there is shown what happens when the graph of the original data is shown with the variables employed, as in (e), with the state of fornix on the horizontal axis. This looks a little like a crossing interaction. For the transected group, "yes" is clearly higher but for the intact group "no" is a little higher. However, it is still a spreading interaction because it is not crossing on both of the representations. Also, the difference for intact animals was very small. Finally, we can see in (d) that we can make the spread go from left to right by changing the position on the horizontal axis of the "yes" and "no" levels of the previous exposure variable. As a matter of fact, still more changes in appearance may be obtained if the direction of the vertical axis is changed. That is, percent wrong responses could be plotted instead of percent correct responses. If that were done, any line that went up from left to right would now go down. Still, a crossing interaction will have crossing lines no matter how the data are graphed. A spreading interaction will still spread out to the left or right no matter how the data are plotted and will not have crossing lines for at least one of the two uses of the horizontal axis.

In summary, when there are two independent variables in an experiment, the main effect of each of the variables may be isolated, as well as the interaction between the two variables. Three main kinds of interaction may be obtained: zero (or no) interaction, spreading interaction, and crossing interaction.

FACTORIAL DESIGN IN TESTING SINGLE-RELATION HYPOTHESES

We will now consider the factorial experiment as used in the testing of single-relation hypotheses. These are the experimental hypotheses that have been discussed in earlier chapters. The relation is hypothesized between a single independent variable and a dependent variable; for example, state of fornix (transected or intact) and recognition-memory performance.

Removing Associative Confounding of Underlying Variables

You will recall that Gaffan was able to show that transection of the fornix affected memory, not registration, and that it was specifically recognition memory that was affected. In each case, he *attributed* the effect to an *underlying* variable, memory and recognition. The problem he solved was that of confounding. We have seen in Chapter 5 that confounding by a second variable may be eliminated by means of a control condition for comparison with the active treatment. However, that remedy is not available when the experiment is done to attribute an underlying variable to the independent variable rather than simply to determine whether or not it had an effect on behavior.

Control of Natural Confounding   In order to test for memory, Gaffan had to employ a long interval (130 sec) after the sample trial. Yet, if he had compared the fornix-transected group and the fornix-intact group at just the long interval (130 sec), after the sample trial he would not have been able to tell whether any difference found was due to effect on memory or effect on registration. Since the underlying variable of interest was memory, the confounding was with the secondary underlying variable, registration. Should memory difference or registration difference be attributed to state of fornix? Let us diagram the situation if only a long delay had been used, showing the *possible* effects of transection of the fornix:

		State of Fornix	
		Transected	Intact
Delay Interval			
Long	Memory (required)	Possible	Unaffected
	Registration (required)	Possible	Unaffected

We see in the comparison across the long delay interval, the difference between transected and intact groups could be due to either *potential* underlying variable, as both are required to perform the task. Hence, if the difference is attributed to memory, there is confounding by registration, the other underlying variable.

In Chapter 5, the confounding between state of fornix and state of surrounding area was removed by the control condition of damaging the surrounding area of fornix-intact animals. The present confounding cannot be removed by any control condition for the fornix-intact group. That would require putting the same possibility of loss of registration into the fornix-intact animals as exists for the fornix-transected animals. This cannot be done, because the possibility for the fornix-transected group is unknown. Hence a different solution is required: to introduce a level of *delay interval* in which there can be no effect attributable to memory. A short delay between sample and matching will do just that. Memory is not required for good performance. Here, the *possible* effects of fornix transection are shown as:

		State of Fornix	
		Transected	Intact
Delay Interval			
Short	Memory (not required)	Not possible	Unaffected
	Registration (required)	Possible	Unaffected

Now the comparison between groups involves only the possible effect of registration. There is no confounding. Still, all that could be learned here is whether state of fornix is related to registration, which was not the purpose of the experiment.

However, use of delay interval as a *second independent variable*, in addition to state of fornix, will allow us to find whether or not transection of the fornix interferes with memory. That is to look at the two foregoing diagrams as making up one factorial experiment. If we go back to Figure 8.1, we can see that there was no effect of the transection on registration, since the two groups performed the same at the 10-sec delay. The spreading interaction between state of fornix and delay interval shows that the effect was on memory. The confounding of memory and registration is an example of *natural* associative confounding as described in Chapter 5. Memory cannot occur without previous registration of what to remember; it is in the "nature of things," true in the outside world just as in the experiment. The example was developed for only two levels of delay solely for reasons of simplicity. As we saw in Chapter 7, multilevel independent variables are always more informative.

Control of Artifactual Confounding   But what kind of memory was affected? The particular kind that inspired the research was failure of people with damage to the hippocampal area to recognize objects after seeing them. This kind of deficit can easily be established with people. All you have to do is show them some small objects such as Gaffan used in his experiment and then later find whether they can recognize whether they have seen these objects when they are shown a mixture of old and new ones. We have already seen that Sternberg (1969) was able to use a recognition task with people as subjects: They were to press one button if the probe item was one of the positive set previously shown and the other button if it was not.

But you can't chat with a monkey. (You *could* chatter.) What Gaffan did was allow the monkey to associate a certain object with the Sugar Puff reward so that the monkey had a chance to demonstrate its ability to recognize. The recognition task introduced two possible differences between the transected and intact animals, in recognizing the positive objects and in making the association between the positive objects and the Sugar Puff reward. The transected animals could simply have been deficient in ability to associate. You will recall that the negative objects were all new ones, not previously presented. Let us look at a diagram of how an animal might be affected in the performance on the *recognition* task:

		State of Fornix	
		Transected	Intact
Previous Exposure to Negative Objects			
No	Recognition (required)	Possible	Unaffected
	Association (required)	Possible	Unaffected

The same strategy was employed to remove the confounding as in the subexperiment to distinguish between memory and registration.

We have already seen association tested both in people and dancing mice. Calfee and Anderson's subjects (1971) associated numerals with trigrams. Yerkes and Dodson's subjects (1908) associated appearance of tunnel with shock. An association task was found by Gaffan (1974) that did *not* require the animals to distinguish between previously seen and new objects, i.e., to recognize. All of the objects were previously seen; some of them were positively associated with the reward and the others were negatively associated. Thus, the possible effects of fornix transection on the *association* task may be represented as follows:

		State of Fornix	
		Transected	Intact
Previous Exposure to Negative Objects			
Yes	Recognition (not required)	Not possible	Unaffected
	Association (required)	Possible	Unaffected

If that were the only task used, all that could be found would be whether transection of the fornix had an effect on association. However, the underlying variable of interest was recognition. By using *both* the recognition and association tasks, a second independent variable was introduced, previous exposure to negative objects: yes or no. You see in Figure 8.2 that the fornix-transected group was not at a disadvantage as compared with the intact group when only association was required, but it was at a great disadvantage when recognition of objects as old or new was also required. The spreading interaction between state of fornix and previous exposure to negative objects allowed identification of the underlying variable that differed for the two groups of animals; it was recognition, not association.

As in the distinction between memory and registration, there was no way in which a control *condition* could have been used to produce the same possible effect on association in the intact-fornix group as in the fornix-transected group. A factorial experiment was necessary, with the second independent variable being previous exposure to negative objects. That is, this control *variable* was needed. The confounding between recognition and association was entirely due to the need of the experimenter for the monkeys to make association between object and Sugar Puffs so that recognition could be tested. In the real world of people and animals, there are recognitions that take place without such associations with rewards. Hence, the confounding was an *artifactual* one, introduced by the requirements of the experiment. For distinguishing between memory and registration, there was control over *natural* confounding. Nothing can be remembered if it is not first registered. Still, the strategy of control was the same.

In the condition needed for testing the hypothesized underlying variable, it was confounded with a second underlying variable. A new condition was introduced, in which only the second underlying variable was necessarily involved in any different effects between levels of the primary independent variable. Thus, the two conditions for testing made up a control *independent variable*, with a more active and a less active level.

EXPERIMENT WITH CONTROL FOR CONFOUNDING OF UNDERLYING VARIABLES

	Primary Independent Variable	
	Active level	Inactive level
Control Independent Variable		
More active level		
Hypothesized underlying variable	Possible effect	Unaffected
Confounding underlying variable	Possible effect	Unaffected
Less active level		
Hypothesized underlying variable	No possible effect	Unaffected
Confounding underlying variable	Possible effect	Unaffected

In the "more active" level of the control variable, it is possible that either the hypothesized underlying variable (memory, recognition) or the confounding one (registration, association) is affected by the active level of the primary independent variable (fornix-transected). In the "less active" level of the control variable, the hypothesized underlying variable is made inactive for the active level of the primary independent variable. The spreading interaction of the primary and control independent variables allows testing of the hypothesis of the underlying effective process.

This control benefits *internal* validity. In the pristine variety of the ideal experiment, it would be possible to have memory without registration; it would be possible to assess recognition in monkeys without using an association with Sugar Puffs. The factorial experiment, with the second independent variable functioning as a control, allows the testing of hypotheses to approach the ideal experiment.

Control for Generalization

We shall now go back to another experiment introduced in Chapter 5, that by Roy Wise and Vivien Dawson (1974) on diazepam-induced eating. The hypothesis was tested that this drug has a direct effect on eating—it makes the animals hungry. In the experiment, it was convenient to put each animal in a test cage for 15 minutes and to measure how much food it had eaten. This introduces a problem in testing the hypothesis. There is an alternative hypothesis of why a rat might eat more in the test cage with diazepam injection than with the control injection. It is called the *antianxiety hypothesis*. The idea is that the novel situation tends to make rats anxious and so inhibits their eating. Then the effect of diazepam—a known tranquilizer—would be to relieve the anxiety and so the rat would eat normally. Obviously, a control is in order.

An experiment was performed on 15 rats using a design with across-subjects counterbalancing. Injections were given 15 minutes before testing. "In the feeding tests 20 grams ($\pm$.5 grams) of preweighed food was placed in the cage or test box before testing; the uneaten food and crumbs were reweighed after testing" (Wise and Dawson, 1974, pp. 931–932). The comparison of diazepam-injected and placebo-injected treatments is shown in Figure 8.4. It can be seen that the rats ate much more with diazepam than with placebo. Under each treatment, somewhat more was eaten in the home cage than in the test cage. The interaction was near zero.

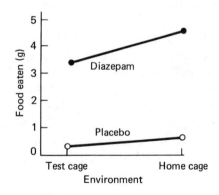

Figure 8.4 Food eaten, with and without diazepam, in the home cage and in the test cage.

The noteworthy finding is that there was as much difference between diazepam and placebo for the home cage as for the test cage, perhaps a little more. The effect of diazepam was direct and was not dependent on the reduction of anxiety. That is not to say that there was no indication of more eating with less anxiety. Still, the main effect of environment (test cage or home cage) was only .7 (grams) as compared with the main effect, introduction of diazepam: yes or no, which was 3.5 (grams).

This was only one of the many subexperiments in the large study by Wise and Dawson (1974). However, the central focus, overall, was testing the hypothesis of a specific, direct effect of diazepam on eating behavior. In the particular subexperiment just described, the alternative hypothesis was that the effect was not simple but rather combinational, involving two variables: introduction of diazepam and anxiety-producing environment. Actually, it was not necessary to do a factorial experiment to compare these two hypotheses. If the testing had been done entirely in the home cages, the antianxiety hypothesis would have been tested adequately. We can see this in the diagrams that follow. First, consider use of the test cages:

	Introduction of Diazepam	
	Yes	No
	Active	Inactive
Environment Anxiety-Producing		
Test cage, yes	Active	Active

Diazepam is, of course, the "active" level of introduction of diazepam. The test cage is anxiety-producing and is thus indicated as "active" for both "yes" and "no" treatments. When more eating is found with diazepam, it could simply be due to that variable or to the different *combinations* of the diazepam and environment variables. With no diazepam, the anxiety in the test cage would prevent eating. With diazepam, there would be action against anxiety, and so the animal would eat freely. This possibility of such a combinational effect is eliminated by testing in the home cages:

	Introduction of Diazepam	
	Yes	No
	Active	Inactive
Environment Anxiety-Producing		
Home cage, no	Inactive	Inactive

Now there is no further combination of the active level of two variables. Any difference between the two diazepam levels must be due to a direct effect.

Why did the experiment include the test cage environment if it was not really necessary? Why was a factorial experiment done? The answer is that there can always be criticism of an experiment because of the level of another variable that has been held constant. The claim that the experimental hypothesis holds *generally* is disputed. It is argued instead that the effect depended on the particular *combinations* compared and so cannot be generalized. The antianxiety hypothesis was that it is not true that diazepam has the simple effect of increasing hunger and that this would generally be shown, regardless of other circumstances such as environment. The simple effect hypothesis, then, is one that generalizes to other circumstances. Whenever it is possible to do an experiment using more than one level of another variable, there is increased safety in the claim that the effect is simple and generalizable, i.e., not dependent on the level of other variables.

We are now ready to answer a question asked in Chapter 5 in respect to the experiment on the work ethic in which lever pressing for reward was preferred over freeloading: Why *Indian girls*? If only one type of child had been used, the claim could be made that lever pressing will be preferred only when combined with this type. Specifically, it has been suggested that the preference shown in a previous study by one of the same investigators (Singh, 1970) was because the children had been imbued with the white Protestant ethic that "one should work to obtain rewards" (Singh and Query, 1971, p. 77). To control for the possibility of this peculiar fetish, Singh and Query then did the experiment that not only included white girls but also Indian girls, Indian boys, and white boys. The second variable in the experiment might be called *cultural background*. Since the effect did not depend on cultural background, the simple hypothesis may be generalized more confidently that people prefer active effort over freeloading. As a matter of fact, it is not just people. The same result has been found for pigeons and rats (Jensen, 1963; Neuringer, 1969; Carder and Berkowitz, 1970; Singh, 1970). However, it must in all honesty be reported that cats seem to be different. Kenneth Koffer and Grant Coulson (1971) report this finding in their article, "Feline indolence: Cats prefer free to response-produced food." Six tomcats were studied (Zero, PeeWee, Golem, Nemo, Ivan, and Selma) and each ate the freely available Puss 'n' Boots Beef Cat Food before starting to press a bar to obtain the same type of food. This is not the first suggestion that cats have their quirks (or maybe they alone are completely rational).

The use of different levels of a second variable, including the subject variable, is for the purpose of establishing generality, not for eliminating confounding. It is an approach to the *completely appropriate experiment*, that perfect experiment in which the experimental hypothesis would be tested under all levels of other variables, with all subjects to whom the results are applied. Consequently, its use represents an improvement in *external* validity.

In Chapter 5, generalization was treated in terms of *internal* validity. To have a generalizable relation between an independent and dependent variable, we must make sure that we have the kind of independent variable that is pure and not confounded with other variables. This is the only kind of independent variable that could possibly function in situations that are superficially very different. Now we see that this is not sufficient for safe generalization. If the independent variable shows a given relation with a dependent variable only at certain levels of another variable, it must be affecting behavior indirectly, through a combinational effect, not a simple one. The generalization of an effect across levels of another variable is a matter of *external* validity. To boil down what has just been said: Generalization is achieved through obtaining *simple* effects of *isolated* variables.

Counterbalanced Designs

You will find in reading experimental articles that counterbalanced designs are often treated in factorial terms as including a second independent variable even though the hypothesis concerns only one independent variable of interest. For example, in each of the pair of experiments by Kennedy and Landesman (1963), the relation was sought between height of working surface and the number of blocks turned. Each height occurred equally often across subjects on each of the six successive trials. A mean can be obtained for performance *on each trial. Positional order* has automatically become a second independent variable. Thus, we can find the separate main effects of height of working surface and of positional order. We can also find the interaction between the two independent variables. Sometimes the interaction does turn out to be large, and some interpretation is demanded. However, even when that is not the case, the positional order variable is an automatic feature of counterbalanced designs (other than reverse counterbalancing in a multilevel experiment). Such designs, then, should always be analyzed factorially.

TESTING COMBINATIONAL HYPOTHESES

Here experimenters move a step up in ambition. They attempt to test experimental hypotheses of how two independent variables work in combination to affect behavior. There are many cases in which it is known that more than one independent variable will affect a behavior. Even before the Yerkes and Dodson study (1908), it was certain that both strength of shock and difficulty of discrimination will be related to number of trials required for learning. In the memory search task, Sternberg (1969) knew in advance that a longer time would be required as the number of items in the positive set was increased. But he also knew that the time would be lengthened if the probe item was made difficult to see. You will recall from Chapter 7 that Yerkes and Dodson had reasons for hypothesizing a particular *combinational* relation between strength of shock and difficulty of discrimination on the one hand and number of trials to reach the criterion of learning on the other. We shall see that Sternberg used a model of how information is processed in recognition to hypothesize the combinational relation that should be found between his two independent variables and reaction time for recognition. In both experiments, the experimental hypothesis was concerned with the *interaction* between the variables. Two new experiments also will be introduced, so that each of the major types of interaction described previously in this chapter will be included as an experimental hypothesis.

The Yerkes-Dodson Law: Location of Optimum Level

A preview was given in Chapter 7 of the factorial experiment performed so many years ago on the relation between strength of shock and the learning of a black-white discrimination by dancing mice. Yerkes and Dodson repeated their procedure using both a more difficult discrimination and a less difficult discrimination. The discrimination was made easier by reducing the light entering the black tunnel, making it even blacker. The discrimination was made more difficult by reducing the light entering the white tunnel, so it looked more like the black one.

A separate group of mice was now tested at each of five levels of shock for the easy discrimination and at each of four levels of shock for the difficult discrimination. For each group, the mean number of trials was found for reaching the criterion of learning. The entire results, including those previously presented in Figure 7.3 for the medium discrimination, are shown in Figure 8.5. The main effect, difficulty of discrimination, is obviously a strong one. Generally, across all levels of shock, there was quicker learning as the tunnels were easier to tell apart. There is perhaps also some overall superiority for stronger shocks, as the means toward the right tend to be a little lower than those toward the left. However, of real interest is the interaction between the two independent variables, strength of shock and difficulty of discrimination. The hypothesis stated in Chapter 7 was supported. We have already seen that the minimum number of trials required for learning the medium discrimination occurred at the shock level of 300 units. Now, it is also seen that for the difficult discrimination, the minimum occurred at a weaker shock, 195 stimulation units. Finally, it is seen that for the easy discrimination performance was still improving at 420 stimulus units. It is quite possible that the mice would have done still better with an even stronger shock.

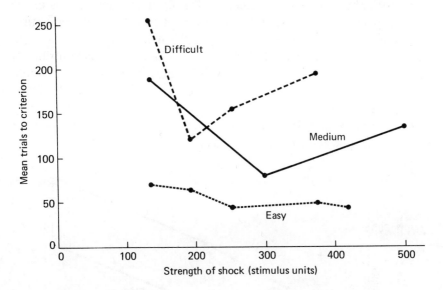

Figure 8.5 Relation between strength of shock and number of trials for learning discrimination of different levels of difficulty, by dancing mice.

Memory Search Revisited: Variables that Do Not Interact

You will recall the experiment of Chapter 7 in which Sternberg (1969) measured the time it took the subjects to search through a set of numerals held in memory to tell whether a probe numeral was one of them or not. He found an absolute-absolute relation. For each increase of an item in the memory set, there was an equal lengthening of time required for recognition, about 35 msec. This happens when the probe stimulus can be seen very clearly. What should happen when the probe stimulus is made difficult to see very clearly—e.g., poorly printed? We know that this will lengthen time for recognition. But will there be the same lengthening no matter what the size of the positive set that is being tested? Perhaps with a large set there will be a much greater effect of lack of clarity. Sternberg hypothesized that equal lengthening would occur— zero interaction—instead of more lengthening as set size increased, i.e., spreading interaction.

In order to make the probe numeral unclear, he projected on it a checkerboard pattern. It could still be made out, but with considerably more difficulty. The reaction times for the different combinations of the two independent variables, size of positive set and clarity of probe stimulus, are shown in Figure 8.6. Each of the main effects is seen to be strong. There was an overall 115-msec lengthening in reaction time in going from 1 to 4 items in memory. This is only a shade more than would be expected if each new item lengthened memory search by 35 msec, giving a total lengthening of 105 msec with an increase of three items. Overall reaction time is seen to be lengthened about 70 msec when the probe numeral was made to be unclear.

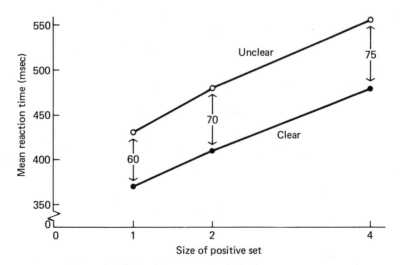

Figure 8.6 Relation between size of positive set and reaction time to clear and unclear probe stimuli.

In contrast, there was only a negligible amount of interaction. With Set Size 1, the difference between the clear and unclear probes was 60 msec. With Set Size 4, it was 75 msec. At most, the interaction amounts to 15 msec. This could easily have occurred by chance and is very small compared with the two main effects. In summary, Sternberg predicted zero interaction between the two independent variables he knew would affect search time and obtained a result that supported his hypothesis.

Early Experience and Problem Solving: Unequal Effects

It is always easy to second-guess. After an experiment has been done, and some unexpected result obtained, one can usually tell the investigators that this is the result they should have hypothesized and why. Many of us, including this writer, are gifted with 20/20 hindsight. This should be taken into account in the discussion that follows of an experiment on early experience and problem solving in rats by Victor Denenberg and John Morton (1962).

There were two independent variables in their study. The first was preweaning handling. During the 24 days following birth, in which the pups were being weaned, half of the pups were handled once a day, and half were left alone with their mother and litter mates. "Handling consisted of removing the pups from the cage and placing them individually into cans partially filled with shavings for 3 minutes, after which they were returned to the home cage" (p. 1096).

There were three subexperiments, in which the other independent variable, postweaning environment, was arranged somewhat differently. Denenberg and Morton's description for Experiment 3—the one to be discussed—is as follows: "At weaning (25 days) a number of handled and nonhandled litters were placed in 7-in. by 9½-in. by 7-in. laboratory cages where they remained until adult testing. Other handled and nonhandled litters were placed in free environment boxes. Four free environment boxes were available; two boxes contained handled litters and the other two boxes contained nonhandled litters" (p. 1097). The two levels of postweaning environment, then, were cages and free environment. The cages were curtained off, making for a very restricted environment. The free environment box was 4 feet square "and contained platforms, tunnels, ramps, and alleyways" (p. 1096). All this made for an enriched environment. Rats lived either in cages or in free environment boxes until they were mature (50 days of age).

Then came the maze testing. The Hebb-Williams maze used can be set up in a number of different ways, in which the path to obtain food is different. Each layout is called a *problem*. Rats were run one at a time, first on practice problems for 15 days and then on the scored test problems for 12 days, one problem per day. "The *S*'s score was the total number of errors made throughout the testing period" (p. 1097). The dependent variable was the mean number of errors for the subgroup to which the animals belonged: handled, cages; handled, free environment; nonhandled, cages; or nonhandled, free environment.

The unlooked-for results are shown in Figure 8.7. Obviously the main effect, preweaning handling, does not amount to much, the mean performance for handled and nonhandled being about the same. The other main effect, postweaning environment, seems somewhat larger, as free environment was lower in errors than cages both for handled and nonhandled animals. However, the difference was not statistically significant. What catches one's eye is the very marked spreading interaction between preweaning handling and postweaning environment. For handled animals, the difference in mean error for the two environments was only 14, while for nonhandled animals it was 83—an interaction of 69. For environment, *unequal effects* were found: small for handled animals, large for nonhandled animals.

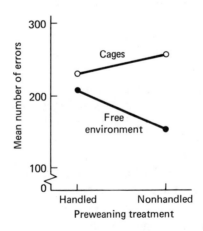

Figure 8.7 Effect of preweaning handling and postweaning environment on errors during maze learning by rats.

Here is the reason such an interaction might have been hypothesized. It is known that preweaning handling will make rats less emotional or wild. This might also be described as more dependent or less adventurous. The free environment would make an animal a problem solver only if the animal took advantage of its opportunity. Thus, this environment helped nonhandled animals, which had not been made dependent by human care, but could not help the handled animals.

However, the two investigators were not interested in trying to account for the interaction. They concluded rather flatly that the preweaning experience has no effect on later problem solving. Their statement about the interaction was that it "can reasonably be attributed to chance fluctuation" (p. 1097). As was discussed in Chapter 6, the whole purpose of a test of statistical significance (and here the probability of the null hypothesis was less than 1 in a 100) is to have a standard for either accepting a result or attributing it to chance fluctuation.

An Experiment on Irrelevant Information: Compatibility

As we have seen, experimenters sometimes get results they do not expect. Less often, the findings make such clear sense that they amount to a discovery. The name *serendipity* has been given to the knack of finding one thing while you are looking for another. It is highly recommended. Such was the experience of J. Richard Simon and Alan Rudell (1967). The effect they discovered will seem to you to be obvious and something they should have expected from the beginning. However, this very simplicity really means that a very uncluttered picture has been shown of the relation between independent and dependent variables. It could not, in fact, have been quite that obvious prior to their experiment. After all, their investigation was in the field of reaction time, in which experimentation had been going on for 100 years, and no one else had reported that finding.

What Simon and Rudell had hypothesized was that reaction time should be shorter when the signal is given to the dominant hemisphere of the brain. This is the left hemisphere for right-handed persons and the right hemisphere for left-handed persons.

Their hypothesis was that a tone signal to the right ear (which is connected with the left hemisphere) will be responded to more quickly by a right-handed person than will a tone to the left ear (which is connected with the right hemisphere). Further, the opposite relation will hold for left-handed persons: shorter reaction time to a tone to the left ear. The investigators went on to devise a splendid experiment to test the hypotheses. Subjects were all college students: 16 female left-handers, 16 female right-handers, 16 male left-handers, and 16 male right-handers.

The task was to follow the command given in one earphone or the other: to press a button with the left hand when the command was "left" and to press a different button with the right hand when the command was "right." This was to be done regardless of the earphone issuing the command. On a block of 132 trials, the commands "left" or "right" occurred equally often at random, and they both occurred equally often at random in the two ears. The expectation of Simon and Rudell was that, overall, reaction times would be shorter for right-handed subjects on trials in which the command entered the right ear than when it entered the left ear, and vice versa for the left-handed subjects.

The results that test this hypothesis are shown in Figure 8.8(a). As is evident, the prediction did not hold. For all combinations of handedness and ear stimulated, the reaction time was much the same, between 404 and 410 msec. The possible spreading interaction is so slight (7 msec) that it may be safely disregarded.

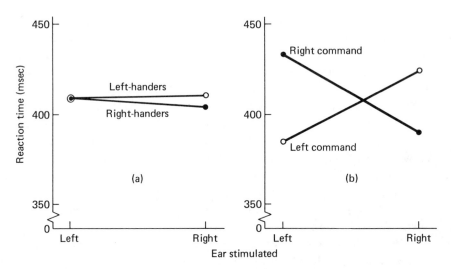

Figure 8.8 Reaction time to commands given in the left and right ears: (a) for left- and right-handed subjects; (b) for commands to respond with the left or the right hand.

On the contrary, as is shown in (b), there was a huge interaction between ear stimulated and command: left or right. This, of course is a *crossing* interaction; it amounts to 85 msec. As was already evident in (a), there was scarcely any main effect for ear stimulated (5 msec). Similarly, the main effect, command, amounts to only 3 msec.

It should be pointed out that subjects were instructed to pay no attention to the ear in which they heard the command but only to the word spoken, *left* or *right*. Obviously, they were unable to disregard the *irrelevant* information.

The Advances Provided

In the four studies just described, the use of a second independent variable allowed the testing of experimental hypotheses that could not be tested with a single independent variable. To be sure, not all of these hypotheses were thought of in advance by the experimenters. Still, the information would not have been available at all without the second variable.

It was only by using different levels of shock that the theory of two underlying processes in black-white discrimination learning could be tested. The experimental hypothesis was tested that as difficulty increases the shock that will lead to a minimum—fastest learning—will become lower. That prediction was based on the greater demand on the differentiation process with more difficult discriminations. As shock level is decreased, the dancing mouse was more "careful and deliberate." Even though there may have been poorer association of lightness of tunnel with shock—the second underlying process in the theory—it was more than made up for by the better differentiation. If the animal is too harassed to notice white or black, it has nothing to associate with shock. The Yerkes-Dodson law has provided an insight in respect to motivation that has remained vital for a very long time.

Each of the other three studies is an example of one of the main classes of interaction based on curves between independent and dependent variables that go only in one direction: i.e., that do not have a maximum or minimum. As was described early in the chapter, *zero* interaction is shown by parallel lines for different levels of a second independent variable. Where the lines spread out to the right or come together, the term *spreading interaction* is used. Where the lines cross, no matter which variable is put on the horizontal axis, the term *crossing interaction* is used.

In Sternberg's memory search experiment, the interaction between number of items in the positive set and clarity of probe stimulus was found to be close to zero. While the line for the less clear probes was higher than that for the clear probe, it was essentially parallel with it. An experimental hypothesis of zero interaction follows from Sternberg's model of the processes in memory search. We have already shown that the straight-line relation between number of items and reaction time supports the idea of scanning each of the items in turn to determine whether there is a match. According to his model, all of that takes place at a later *stage* of processing than that at which the probe is identified, the "stimulus-encoding" stage. A probe stimulus that is harder to identify just adds a constant amount to the reaction time, no matter how many items are in memory. If it made the scanning more difficult, then it would have added more time when the number of items in memory was larger. Thus, if lack of clarity affected the same stage of information processing as did number of items, there would have been a spreading interaction—a larger difference between a clear and an unclear probe with a larger positive set.

If we are to accept the spreading interaction found in Denenberg and Morton's experiments (1962) on pre- and postweaning effects on maze learning (which the investigators themselves do not!), we may have obtained an insight into the interplay of early factors on later behavior. It is that a free environment will not necessarily be helpful in making an animal a good learner. This will not happen with the wrong preweaning experience. All the tiny pups with preweaning handling had to do to see some more of the world was to be carried around by the experimenter. Maybe they were just waiting for a lift during their time in the free-environment boxes.

Now we come to the "simple-minded" compatibility study. For fastest responses, "left" commands should go in the left ear and "right" commands in the right ear. With the incompatible relation, there seems to be an interference that cannot be disregarded. But why is this? Simon and Rudell (1967) say that the tendency to respond in the same direction as the tone is a "population stereotype" (p. 300). That makes it sound like a learned habit. In the United States, you flip a switch up to turn on the light; in England, you flip it down. We can see how that would lead to a learned habit that would make for trans-Atlantic slip-ups. It is difficult to see how this could have been learned for the ear-hand relation. Simon and Rudell also state: "These results suggest the existence of a strong natural tendency to associate right-ear stimulation with a right-hand response and left-ear stimulation with a left-hand response" (p. 303). That would seem to mean that it is *not* learned. Despite many subsequent experiments on this topic, the nature of the underlying processes remains obscure. Psychology advances by turning up interesting questions quite as much as by providing clever answers.

In all of the discussion of this chapter, we did not get beyond two independent variables. Of course if a counterbalanced design is used, such as the across-subjects Latin squares in the Kennedy and Landesman study (1963), then there is an automatic second variable—positional order. If such a design is used with two independent variables of interest (suppose size of block to be turned was varied as well as height of working surface), then there would technically be three independent variables: height of working surface, size of block, and positional order of trial. Thus, in the literature you will see three-factor designs for studying the combinational effect of two independent variables of interest.

However, there are also experiments in which more than two independent variables of interest are, in fact, used. When there are just the two independent variables, the interaction between them is called a *first-order interaction*. When there are three independent variables, there is also a *second-order interaction*. For example, in a reaction time study that formed part of the investigation already discussed, Sternberg varied: (1) number of alternatives, (2) clarity of stimulus, and (3) compatibility between stimulus and required response. This gave him three first-order interactions between these independent variables: number × clarity, number × compatibility, and clarity × compatibility. The second-order interaction is number × clarity × compatibility. A statement of what he found is that there is the same spreading interaction between number and compatibility for clear and unclear stimuli. Use of this higher-order interaction permitted the testing of a more detailed model than was previously discussed. Let's just stop right now, with three independent variables!

CLASSIFICATION OF EXPERIMENTAL DESIGNS

We are now ready to look back at the experimental designs covered in this book. There will be no others added in the last chapter, since it deals with correlational studies, not those in which the experimenter manipulates independent variables. The various experiments discussed differ from one another along three basic dimensions. These correspond to distinctions that were made in Chapter 4, Chapter 7, and the present chapter.

In the very first studies, where only a single subject was used, the control for internal validity was to use either an alternating, random, or counterbalanced order of trials. Even if a number of subjects are used in an expansion of this kind of design, the controls remain within-subject controls. Therefore, a within-subject comparison is made between treatments or levels. When internal validity was controlled by having different groups of subjects for the different treatments or levels, the comparison made of the effect of the independent variable was between the groups of subjects. In the last chapter, internal validity was controlled by testing the subjects with counterbalanced sequences of levels of the independent variable. All comparisons between levels were thus made across subjects. The first dimension for classifying experimental designs may be called *basis of comparison: within-subject, between-groups, or across-subjects.*

Also in the preceding chapter a distinction was made between those experiments in which there were either only two treatments or else several qualitatively different treatments and those in which there was quantitative grading of the independent variable. That has been discussed so recently that it should still be fresh in your memory. The second basic dimension of experimental designs is *gradation of independent variable: ungraded and graded.*

The third distinction among experimental designs was introduced in this chapter. It is whether there has been one independent variable used or more. Thus, the third basic dimension is *number of independent variables: one variable or factorial.*

Since there are three bases of comparison, two types of gradation, and two possibilities for number of independent variables, there would seem to be exactly 12 different pigeonholes into which the design for any experiment may be placed. A number of the experiments described in this book can be so categorized. Examples are given in Table 8.2. Thus, in the experiment on early experience by Denenberg and Morton (1962), between-groups comparisons were made for each independent variable, preweaning treatment and postweaning environment, and both were ungraded. This experiment fits neatly in Pigeonhole Number 8. However, there are factorial experiments that do not fit in so well. The two or more independent variables may differ in respect to basis of comparison or to gradation of independent variable. Thus, there are many more than 12 possible combinations. In Gaffan's experiment in which monkeys were tested at different time intervals after their response to the sample, one independent variable, state of fornix, naturally used a between-groups basis of comparison, whereas the other independent variable, time interval, used across-subjects comparisons. Again, state of fornix was ungraded (just transected or intact) while time interval was somewhat graded (three levels).

TABLE 8.2
The 12 Homogeneous Experimental Designs

Number of Independent Variables	Within Subject	Basis of Comparison Between Groups	Across Subjects
One Variable			
Ungraded	1 Memorizing Piano Pieces (Chapter 1)	2 Mental Practice (Chapter 4)	3
Graded	4 Subjective Heaviness (Chapter 7)	5 Presentation Rate (Chapter 5)	6 Reaction Time (Gottsdanker and Way), fixed interval (Chapter 7)
Factorial			
Ungraded	7 Fornix Transection Delays (Chapter 8)	8 Early Experience (Chapter 8)	9 Diazepam (Chapter 8)
Graded	10	11 Yerkes-Dodson Law (Chapter 8)	12 Block Turning (Chapter 7)

When basis of comparison and gradation are the same for both independent variables, as in the experiment of Denenberg and Morton, the design is called *homogeneous*. When one or the other differs for the independent variables, as in Gaffan's experiment, the design is thus *heterogeneous*. Factorial designs are often referred to by the number of levels for each independent variable. The Denenberg and Morton study (1962) used a "two-by-two" design. Gaffan (1974) used a "two-by-three" design in the subexperiment on delays.

The three-dimensional classification given is in terms of the basic dimensions. You will recall that we discussed three different orders for within-subject comparisons and that there are a number of different ways of constituting groups for comparison, random assignment, stratified random selection, etc. Thus you may read an entire issue of a journal and not find exactly the same design in any two articles.

SUMMARY

It was shown in an investigation to test the hypothesis that damage to the fornix of a monkey affects its recognition memory, that one factorial experiment was needed to establish that the effect was on memory and another to establish that the effect was specifically on recognition. In each of these experiments, a second independent variable was included, in addition to the primary variable of interest, state of fornix. The term *factorial experiment* means that more than one independent variable is manipulated.

In a factorial experiment, a *main effect* is found for each independent variable, and the *interaction* is found between the independent variables. The experiment on the effect of transection of the fornix was used to illustrate the meaning of these terms and to show how they may be measured when there are two independent variables, each having two levels. The main effect of an independent variable is the overall difference between the mean values of the dependent variable for the two levels. The interaction is the difference between two differences: one for each level of the second independent variable.

Graphic representations were given to illustrate the main kinds of interaction that are found in factorial experiments. The way of measuring main effects and interactions on such diagrams was also indicated. One independent variable is scaled on the horizontal axis and a separate line is drawn for each level of the other independent variable, to plot the values of the dependent variable. If the two lines for the other independent variable are parallel, there is *no* (or *zero*) *interaction.* When the lines fan out, either to the right or to the left, there is a *spreading interaction.* When the two lines cross each other, there is a *crossing interaction.*

Factorial experiments are often used in testing single-relation hypotheses, i.e., the effect on behavior of one independent variable. This is the only kind of hypothesis discussed in preceding chapters. Here, a major purpose is in the removal of associative confounding, either natural or artifactual. The factorial experiment provides a control in investigations in which the hypothesis attributes the effect of the independent variable to *an underlying variable.* However, the condition needed for testing the different levels of the independent variable of interest might also differentially influence a second underlying variable. This was illustrated through the two subexperiments in Gaffan's experiment (1974) on the effect of transecting the fornix of monkeys. Confounding of memory with registration was eliminated through use of a short interval between sample and match as well as a long one. There could be no possible effect of memory at the short interval. Confounding of recognition memory with associative memory was removed through use of a pure association task that did not necessarily depend on recognition. In both cases, the second level of the control variable was one at which an effect through the hypothesized underlying variable was *not* possible. A spreading interaction was indicative of the correctness of the attribution of the underlying variable. The factorial design, then, was able to approximate the results of a pristine ideal experiment, in which there was manipulation only of the hypothesized underlying variable. The gain is in internal validity.

External validity may also be improved by use of a factorial experiment. This is the case in which the hypothesis is tested that the independent variable has a simple, direct effect. The alternative hypothesis is that there is a combinational effect in which the active level of the independent variable works only because of a combination with the active level of another variable. Thus, Wise and Dawson (1974) hypothesized that the drug diazepam increases eating directly, by making rats more hungry. The combinational hypothesis was that it acts indirectly by reducing anxiety. Use of an environment that was not anxiety producing—the rats' home cages—gave as much difference between diazepam-injected and placebo-injected animals. In this way, it was shown that there is a general effect of diazepam, not one that depends on level of anxiety. An effect is not generalizable if it depends on the level of other variables. Tests of generalization are made over a subject variable as well as over one that is produced by the experimenter. Indian girls were used in the study of the work ethic in order to generalize the effect over persons with different cultural backgrounds.

Factorial experiments come about automatically when counterbalanced designs are used. In addition to the independent variable of interest, there is also the positional order variable. In across-subjects counterbalancing, the mean is found across subjects for the first trial, second trial, etc.

Of major interest is the use of factorial experiments to test hypotheses that are genuinely combinational. The use of different degrees of difficulty of discrimination was needed in order to test the hypothesis now known as the Yerkes-Dodson law: Optimal strength of shock for learning is lower as discrimination is more difficult. Each of the three kinds of interaction described previously was hypothesized in other factorial experiments. Zero interaction was found by Sternberg (1969) in an experiment in which he varied clarity of probe as well as size of positive set in memory. A spreading interaction was found by Denenberg and Morton (1962) in an experiment in which the two independent variables were preweaning treatment and postweaning environment, with rats as subjects. A crossing interaction was obtained in an experiment on reaction time by Simon and Rudell (1967). The two independent variables were command (right or left) and ear stimulated (right or left). It was found that ear stimulated, an irrelevant variable, greatly influenced reaction time. When the "ear" opposed the command, reaction time was lengthened.

For each of the foregoing experiments, the combinational hypothesis required use of a factorial experiment. They were used to test theories or models that could not be tested with just one independent variable. The Yerkes-Dodson interaction supported a theory based on the underlying variables of stimulus differentiation and association, which are oppositely affected by strength of shock. The Denenberg and Morton results could be understood by the explanation that preweaning handling makes rats unable to take advantage of a free environment to improve ability to learn. The Sternberg results supported his model of separate information-processing stages for stimulus encoding and memory scanning. The finding of compatibility between ear stimulated and command remains an intriguing puzzle. It was also indicated that experiments may have more than two independent variables of interest with more complex combinational hypotheses tested. An interaction between two variables is called a *first-order interaction*; that involving three independent variables is called a *second-order interaction* (and so on).

Experimental designs that have been discussed can be classified according to three main dimensions. The first is basis of comparison (within-subject, between-groups, or across-subjects). The second is gradation of independent variable: ungraded or graded. The third is number of independent variables: one variable or factorial. Designs need not be the same (homogeneous) in respect to both independent variables. For example, one independent variable may use between-groups comparison, and the other may use either within-subject or across-subjects comparison.

QUESTIONS

1. In the experiment on the effect of fornix transection, what was the purpose of testing at more than one time interval between sample and match?

2. Give an example of main effects and of interaction in a factorial experiment.

3. How are main effects measured graphically?

4. Describe the graphic appearance of the main types of interaction when each independent variable has two levels.

5. Why is a control variable often necessary in an experiment in which the experimental hypothesis attributes the effect of the independent variable to an underlying variable?

6. Describe the use of more than one level of a second independent variable in relation to external validity.

7. Why was a factorial experiment necessary for testing the Yerkes-Dodson law?

8. What is meant by the statement that Sternberg's model of information processing led to a particular combinational experimental hypothesis?

9. Draw a graph of the experimental results that would have been obtained if preweaning handling had no effect on later maze learning but if a postweaning free environment was helpful as compared with a caged environment.

10. See whether you can think of a reasonable combinational hypothesis that would be supported by a crossing interaction. (Try to stay away from the kind of study performed by Simon and Rudell.)

11. How does the Yerkes and Dodson experiment fit into the system used for classifying experimental designs?

 STATISTICAL SUPPLEMENT

TWO-WAY ANALYSIS OF VARIANCE

With two independent variables, F is used to test the statistical significance of the two main effects and also of the interaction between the variables. The principles are exactly those stated in the previous statistical supplement. Between-groups estimates of population variance are compared with the within-group estimate to find whether the value of the ratio is sufficiently greater than 1 to allow rejection of the null hypothesis.

The within-group estimate is obtained exactly as has been described. A separate between-groups estimate is found for each of the two main effects and for the interaction. Thus, three values of F are computed; each of the values obtained is compared with the required value at the .05 or .01 alpha level as found in Statistical Table 3.

An Experiment with Two Independent Variables

Let us use our four sets of data on reaction time in a new way. Suppose an experiment on reaction time had been conducted in which the two independent variables were type of signal, light or tone, and type of response, simple or choice. The simple response is to press a button whenever the signal comes on. The choice response is to press a left-hand button when the signal is on the left and a right-hand button when the signal is on the right. Going back to the original designations: Treatment A data shall be taken to represent light-simple; Treatment B, tone-simple; Treatment C, light-choice; Treatment D, tone-choice. Four groups of 17 subjects were used. Thus, our mean reaction times for the four different groups may be shown as:

Type of Response	Type of Signal		
	Tone	Light	Mean
Simple	162	185	173.5
Choice	250	265	257.5
Mean	206.0	225.0	215.5

Differences due to response are thus *row differences*, and stimulus differences are *column differences*. Response × Signal interaction is thus row × column interaction (or $R \times C$). There are r rows and c columns; here, 2 and 2.

Mean Square, Within-Groups

As the same four sets of data are those which have been used previously, we can use the earlier computations for finding mean square (MS), within-groups (WG).

$$SS_{WG} = \Sigma x_{r1c1}^2 + \Sigma x_{r1c2}^2 + \Sigma x_{r2c1}^2 + \Sigma x_{r2c2}^2 \quad \text{(Formula 8.1)}$$

Here

$$SS_{WG} = 4306 + 5808 + 5391 + 4673$$

$$= 20,178$$

As you can see, the designations, if not the numbers, are new. Σx_{r1c1}^2 means that the within-group value of x^2 obtained is that for Row 1 (simple) and Column 1 (tone). Similarly, Σx_{r2c2}^2 is that for Row 2 (choice) and Column 2 (light); etc.

We may again use Formula 7.6 for finding mean square, within-groups (since $r \times c = k$).

$$MS_{WG} = \frac{SS_{WG}}{N - rc}$$

Since there are still 68 subjects divided into four groups, as before

$$MS_{WG} = \frac{20,178}{68 - 4}$$

$$= 315$$

Mean Square, Rows

We shall first find sum of squares, rows, and from it we shall find mean square, rows. The difference is found between each row mean and the overall mean, thus:

$$d_{r1} = M_{r1} - M_{\text{tot}}, \qquad d_{r2} = M_{r2} - M_{\text{tot}} \qquad \text{(Formula 8.2)}$$

Here

$$d_{r1} = 173.5 - 215.5 = -42.0, \qquad d_{r2} = 257.5 - 215.5 = +42.0$$

The sum of squares, rows, is the sum of the squares of these d values times the product of the number of cases per group, n, and the number of columns, c.

$$SS_R = nc\ (d_{r1}^2 + d_{r2}^2 \text{, etc., if there are more rows)} \qquad \text{(Formula 8.3)}$$

Here

$$SS_R = 17.2\ (1764.0 + 1764.0)$$

$$= 119,952$$

Degrees of freedom for rows equals their number minus 1.

$$df_R = r - 1 \qquad\qquad \text{(Formula 8.4)}$$

Here

$$df_R = 2 - 1 = 1$$

Again, a mean square, between groups, is found by dividing that sum of squares by the number of degrees of freedom. Thus, for rows

$$MS_R = \frac{SS_R}{df_R} \qquad\qquad \text{(Formula 8.5)}$$

Here

$$MS_R = \frac{119{,}952}{1} = 119{,}952$$

Mean Square, Columns

Exactly parallel statements may be made about columns as about rows. First,

$$d_{c1} = M_{c1} - M_{tot}, \qquad d_{c2} = M_{c2} - M_{tot} \qquad \text{(Formula 8.6)}$$

Here

$$d_{c1} = 206.0 - 215.5 = -9.5, \qquad d_{c2} = 225.0 - 215.5 = +9.5$$

$$SS_C = nr \, (d_{c1}^2 + d_{c2}^2 \text{ , etc., if there are more columns)}$$

Here   (Formula 8.7)

$$SS_C = 17 \cdot 2 \, (90.25 + 90.25)$$

$$= 6{,}137$$

$$df_C = c - 1 \qquad \text{(Formula 8.8)}$$

Here

$$df_C = 2 - 1 = 1$$

$$MS_C = \frac{SS_C}{df_C} \qquad \text{(Formula 8.9)}$$

Here

$$MS_C = \frac{6{,}137}{1} = 6{,}137$$

Mean Square, Rows × Columns

To find the sum of squares, $R \times C$, you first find the difference between each of the subgroup means and the overall mean. You then add up the squares of these differences and multiply this sum times the number of cases per group. Finally, you subtract from this number the sum of squares, rows, and the sum of squares, columns. Let us now go through this, one step at a time.

$$d_{r1c1} = M_{r1c1} - M_{tot}, \quad d_{r1c2} = M_{r1c2} - M_{tot},$$

$$d_{r2c1} = M_{r2c1} - M_{tot}, \quad d_{r2c2} = M_{r2c2} - M_{tot}$$

Here

$$d_{r1c1} = 162.0 - 215.5 = -53.5, \quad d_{r1c2} = 185.0 - 215.5$$

$$= -30.5$$

$$d_{r2c1} = 250.0 - 215.5 = +34.5, \quad d_{r2c2} = 265.0 - 215.5$$

$$= +49.5$$

$$SS_{R \times C} = n \, (d_{r1c1}^2 + d_{r1c2}^2 + d_{r2c1}^2 + d_{r2c2}^2) - SS_R - SS_C$$
$$\text{(Formula 8.10)}$$

(Note: You have already computed the first part in using Equation 7.4.)

$$SS_{R \times C} = 17 \, (2862.25 + 930.25 + 1190.25 + 2450.25)$$
$$- 119{,}952 - 6{,}137$$

$$= 126{,}361 - 119{,}952 - 6{,}137$$

$$= 272$$

Before we can go on to the final step of computing mean square, rows × columns, we must find the number of degrees of freedom for the rows × columns interaction. Remember that we are comparing the differences in one independent variable as affected by differences in the other independent variable. There are $r - 1$ df in the differences along a row and $c - 1$ df in comparing these row differences across columns. Thus, the total df is the product $(r - 1)$ $(c - 1)$. In our present case, since there are two rows and two columns, the rows × columns interaction equals 1.

$$df_{R \times C} = (r - 1) \, (c - 1) \qquad \text{(Formula 8.11)}$$

Here

$$df_{R \times C} = (2 - 1) \, (2 - 1) = 1$$

Mean square, rows × columns, equals sum of squares, rows × columns, divided by the corresponding degrees of freedom.

$$MS_{R\times C} = \frac{SS_{R\times C}}{df_{R\times C}} \qquad \text{(Formula 8.12)}$$

Here

$$MS_{R\times C} = \frac{272}{1}$$

$$= 272$$

Computing the F-Ratio

There are now four estimates of the population variance, $\bar{\sigma}_{\bar{x}}^2$. They are (1) mean square, within-groups; (2) mean square, rows; (3) mean square, columns; and (4) mean square, rows × columns. We shall use the mean square, within-groups, as the denominator in computing the F-ratio in relation to each of the other mean squares. The entry in the denominator is often called the *error term*, as it refers to the unsystematic variation that the experimenter has not been able to control.

$$F_R = \frac{MS_R}{MS_{WG}} \qquad \text{(Formula 8.13)}$$

Here

$$F_R = \frac{119,952}{315} = 380.80$$

Similarly,

$$F_C = \frac{MS_C}{MS_{WG}} \qquad \text{(Formula 8.14)}$$

Here

$$F_C = \frac{6,137}{315} = 19.48$$

Again similarly,

$$F_{R\times C} = \frac{MS_{R\times C}}{MS_{WG}} \qquad \text{(Formula 8.15)}$$

Here

$$F_{R \times C} = \frac{272}{315} = .86$$

Rejecting or Not Rejecting the Null Hypothesis

As in the statistical supplement to Chapter 7, we use Statistical Table 3 for finding the critical values of F. For F_R there is 1 df in the numerator and 64 df in the denominator. The table entry for 1 and 65 df is an F of 7.04 for rejecting the null hypothesis at the .01 level. Obviously our obtained value of 380.80 allows rejection of the null hypothesis at that level. For F_C we have the same combination of df's in numerator and denominator. So our obtained value of 19.48 again allows rejection of the null hypothesis at the .01 alpha level.

For $F_{R \times C}$, also, we look up the entries for 1 and 65 df. Our obtained value of .86 does not allow rejection of the null hypothesis, even at the .05 alpha level. That critical value is seen to be 3.99. An F of less than 1 can happen in the sampling distribution; it is impossible for it to be statistically significant.

Analysis of Variance Table

The analysis of variance may be summarized in the following table. You should note that degrees of freedom are additive in the same way as sums of squares.

**Analysis of Variance, Experiment on
Type of Stimulus, Type of Response, and Reaction Time**

Source of Variance	SS	df	MS	F	p
Response (Rows)	119,952	1	119,952	380.80	< .01
Stimulus (Columns)	6,137	1	6,137	19.48	< .01
Interaction, Response ×					
Stimulus	272	1	272	.86	
Within-Groups	20,178	64	315		
Total	146,539	67			

PROBLEM: Use the data from the problem in the statistical supplement to Chapter 7 to do an analysis of variance and to fill in the analysis of variance table. Again, the data are from six separate groups of subjects. One variable is size of reward; the other variable is difficulty of problem. The Chapter 7 data should be used as follows:

Difficulty of Problem	Size of Reward (low to high)		
	A	B	C
Easy	Level 4	Level 5	Level 6
Hard	Level 3	Level 2	Level 1

Answer:

Source of Variance	SS	df	MS	F	p
Difficulty (Rows)	433.2	1	433.2	46.33	< .01
Reward (Columns)	15.8	2	7.9	.84	
Interaction, Difficulty × Reward	141.8	2	70.9	7.58	< .01
Within-Groups	224.4	24	9.35		
Total	815.2	29			

CORRELATIONAL STUDIES

By now you know very well what an experiment is. Perhaps you have even been convinced that the experiment, despite all problems of control, offers the best way to test evidence on a variable that is hypothesized to affect behavior. In an experiment, of course, the experimenter manipulates the independent variable according to plan and relates differences in the value of the dependent variable to differences in level of the independent variable.

With the knowledge you have accumulated, try to outline a manipulative experiment on each of the three following hypotheses concerning behavior:

1. Good child-rearing practices will generally produce adults who are psychologically well adjusted, while faulty practices will produce maladjusted adults. (Seems obvious.)
2. The oldest child in a family will tend to be the most intelligent, the second oldest will be the next most intelligent, etc. (Seems rather unlikely.)
3. Persons who score high on a particular aptitude test will succeed at the job of inspecting machined parts more often than those who score low. (Not unreasonable.)

If you are to compare the effects of different child-rearing practices, through a manipulative experiment, you are going to have to convince one group of parents to use "good" practices, such as democratic discussions (which may not be too hard), and another group to use "bad" practices, such as mindless screaming. Of course, the two treatment groups must be well equated, as by random selection. Good luck!

However, the task of that experimenter is easy as compared with the task of doing a manipulative experiment on birth order differences. How to begin? You will want to control not only how many children a couple has, but also the spacing between children. Come to think of it, why not also control the sex of each baby? All this fantasy is bringing us close to the futuristic planned world described in such chilling detail by George Orwell in his novel, *Nineteen Eighty-Four*. It still leaves us far from an effective experiment.

The foregoing experiments are possible, *in principle*. However, depending on how you look at it, they are either impractical or unethical. Studies on these hypotheses have, in fact, been done. But they used a *correlational* approach rather than one of manipulative experimentation. By this it is meant that there was no experimental manipulation to produce differences in behavior. Rather, a correlation was found between existing differences. In the first case, well-adjusted and poorly adjusted adults were compared in respect to information about their early family life, obtained through objective records, through interviews that had been done many years before, and through their recollections. We shall see that problems do arise in ascribing differences in adjustment to differences in child-rearing practices, no matter how obviously true the hypothesis seemed. We shall encounter a now familiar villain, confounding with an associated variable. In the second case, a large number of young men, born at about the same time, were all given an intelligence test at the age of 19. Then separate means were found for first-borns, second-borns, etc. It was found that there *was* a correlation between birth order and intelligence, just as had been hypothesized. Although it had sounded unlikely, it is evidently so. The circumstances of this study—mainly the extremely large number of subjects available—made it possible for the investigators to apply *statistical* controls for confounding with other variables. However, as we shall see, such controls are never as convincing as those described in previous chapters, in which experimental manipulations were used.

The other hypothesis, that an aptitude test will allow the selection of good inspectors, cannot be studied by experimentation for a different reason. There is no independent variable whose effect on behavior might be assessed. Rather, there are simply different measures of behavior of the same subject. Individual differences among subjects on the aptitude test are related to their individual differences in later proficiency in inspecting machined parts. The test scores as well as the job scores are very similar to the *dependent variable* of the previous chapters. However, there is no point in using that term when there is no independent variable.

This chapter should be particularly useful to those of you whose interest in psychology lies more in working with the individual person than in laboratory experimentation. Manipulative experiments are concerned with the *similarity* of responses of subjects—who are not necessarily human. Correlational studies, on the other hand, typically are concerned with *differences* among people such as in intelligence, adjustment, or particular traits of personality. The idea of personality is useful only because people *differ* on many traits. Still, research on individual differences is best understood on the basis of manipulative experimentation. We have already seen that some correlational studies could—in principle—be performed as manipulative experiments. New problems of internal validity arise exactly because this is not a practical possibility. We shall be able to see that a correlational study, as well as a manipulative experiment, gains in internal validity as it is able to approximate an ideal experiment.

Your understanding of the chapter will be advantageous to you in reading articles on correlational studies. You will, first of all, be able to tell whether the investigator attempted to control for confounding variables. You will also be able to judge whether the control methods used were effective. You will even be able to set up certain kinds of correlational study yourself. The example of a test-prediction study should be all that you need to perform a similar study if you go on to the statistical supplement and learn to compute a coefficient of correlation.

While reading this chapter, prepare yourself for answering questions in the following areas:

1. What is meant by a correlational study.
2. Why associative confounding is always present in correlational studies.
3. Control methods for this associative confounding.
4. Practical prediction when a coefficient of correlation may be computed.
5. The dimensions in which correlational studies differ from one another.

AN INVESTIGATION ON ANTECEDENTS OF OPTIMAL PSYCHOLOGICAL ADJUSTMENT

You have just read the title of an article coauthored by four investigators: Ellen Siegelman, Jack Block, Jeanne Block, and Anna von der Lippe (1970). Here, *antecedent* refers to a variation in some earlier state of affairs that affects later psychological adjustment. Don't fret too long over the term *optimal* in describing adjustment. As it is used in this study, it does not mean that the investigators were able to locate some persons who had attained perfection in personality functioning. Rather, they found some adults, in their mid-30s, who approached the optimal much more closely than did some others. High- and low-adjustment groups were compared in respect to their family circumstances when they were children.

Really, it is somewhat surprising that a study of this sort was even possible. Suppose you have located the high- and low-adjustment adults. How are you going to obtain any worthwhile evidence on how they were reared and on the more general attitudes of parents that might have affected later adjustment? The circumstances *were* unusual, requiring groundwork that had been laid more than 30 years before the two groups were identified.

In all, 171 adults were evaluated in respect to psychological adjustment. Correlations were then found with the antecedent measures.

The Subjects

The reason that detailed information was available on the early life of so many adults is that they were all subjects in two *longitudinal* studies carried out at the Institute of Child Welfare at the University of California, Berkeley. A longitudinal study is one in which you make periodic observations of subjects over a long period of time. Samples of children born over a period of several years in the late 1920s in the two cities of Berkeley and Oakland, California, provided the sample. Of course, cooperation of the parents was required. While many of the original subjects were unavailable for testing and several had died since the onset of the study, 171 were located and convinced to participate—a considerable achievement.

The Measure of Psychological Adjustment

Most of the now-adult subjects were interviewed by three different psychologists, the remainder by two. After an interview, a psychologist would describe the subject's personal characteristics by sorting a set of "description cards" into nine piles. This was done on the basis of how well the description on a card fit the individual. More particularly, it was according to how *salient* or conspicuous the description was in respect to the subject. Thus, a description might be "finds excuses for own mistakes." If the psychologist was struck by the number of alibis produced by the subject, he would toss that description card into the first pile. If it was a *somewhat* noticeable trait, he would put it in the fourth or fifth pile. If it was not at all conspicuous of the subject, the card would go into the ninth pile.

There were 90 cards in total, from the "California Q set" (Block, 1961). The psychologist was required to put a prescribed number of cards in each pile, with the largest numbers in the middle piles, to make a kind of bell-shaped distribution.

Before the start of the present study (Siegelman et al., 1970), a different group of clinical psychologists, nine in all, had each sorted the deck of description cards, not to describe any particular individual but rather to define *optimal adjustment*. Let us first look at the descriptions that were considered, on the average, to most *positively* define optimal adjustment. The 13 items that were put in the top categories are shown in the left-hand side of Table 9.1. Anybody who could impress an inter-

TABLE 9.1

California Q Items Considered Most Positively and Most Negatively Defining of Optimal Adjustment[a]

	Most Positively Defining		Most Negatively Defining
Item No.	Item	Item No.	Item
35	Has warmth; has the capacity for close relationships; compassionate	45	Has a brittle ego-defense system; has a small reserve of integration; would be disorganized and maladaptive when under stress or trauma
2	Is a genuinely dependable and responsible person		
60	Has insight into own motives and behavior	78	Feels cheated and victimized by life; self-pitying
26	Is productive; gets things done	86	Handles anxiety and conflicts by, in effect, refusing to recognize their presence; repressive or dissociative tendencies
64	Is socially perceptive of a wide range of interpersonal cues		
70	*Behaves* in an ethically consistent manner; is consistent with own personal standards	22	Feels a lack of personal meaning in life
		55	Is self-defeating
96	Values own independence and autonomy	40	Is vulnerable to real or fancied threat, generally fearful
77	Appears straightforward, forthright, candid in dealings with others	48	Keeps people at a distance; avoids close interpersonal relationships
83	Able to see to the heart of important problems	68	Is basically anxious
		37	Is guileful and deceitful, manipulative, opportunistic
51	Genuinely values intellectual and cognitive matters[b]	36	Is subtly negativistic; tends to undermine and obstruct or sabotage
33	Is calm, relaxed in manner	38	Has hostility towards others[d]
17	Behaves in a sympathetic or considerate manner	76	Tends to project his own feelings and motives onto others
3	Has a wide range of interests[c]	97	Is emotionally bland; has flattened affect

[a]From E. Siegelman, J[ack] Block, J[eanne] Block, and A von der Lippe, Antecedents of optimal psychological adjustment, *Journal of Consulting and Clinical Psychology*, 1970, 35 (3), 283–289. Copyright 1970 by the American Psychological Association: Reprinted by permission.
[b]Ability or achievement is not implied here.
[c]Superficiality or depth of interest is irrelevant here.
[d]Basic hostility is intended here: mode of expression is to be indicated by other items.

viewer in all of these ways is surely a paragon of psychological adjustment! The person would get things done, yet be calm and ethical, etc.,
etc. Now, for the items that most *negatively* define optimal adjustment,
we may look in the right-hand side of the table. It would be hard to be
deficient in all of these ways at once. Thus, we cannot very well imagine
a person who is both basically anxious and one who is emotionally flat.

However, for the purpose of the present study, real use was made
only of the positive set of items. Now the extent to which each subject
was *rated highly on the descriptions that positively characterized optimal
adjustment* was computed. Two groups were selected out of the 171 subjects: a high-adjustment group and a low-adjustment group. The 30 percent of the sample with greatest correspondence to high optimal adjustment were in the first group and the 30 percent with least correspondence
to high optimal adjustment were in the other group. "In all, 24 males and
24 females were rated as high optimal adjustment, and 24 males and 24
females were designated as low optimal adjustment" (Siegelman et al.,
1970, p. 285). Amazing! The middle group of subjects did not figure
further in this particular study.

Measures of Antecedent Variables

In a manipulative experiment, the measure of psychological adjustment would correspond to the *dependent* variable. Let us look at the antecedent variables, which would correspond to the *independent variables*.

Early Family Ratings   Some, but not all, of the parents had been
visited by psychologists and social workers when the subjects were 21 to
36 months of age. They observed the parents "interacting with the
child." Average ratings were made on a number of items, including
marital adjustment and the apparent irritability of both the mother and
father.

Ratings of Mothers   A psychologist interviewed all of the mothers
from two to four times, when the subjects were about 1 year of age to 5
years of age. Ratings were made on both intellectual and emotional characteristics.

Environmental Reminiscences   The present subjects were given an
intensive interview by a psychologist. In this they were "encouraged to
reminisce about the characteristics of their parents, their family life, and
their general childhood environment" (Siegelman et al., 1970, p. 285).
After each interview, the psychologist sorted a set of 93 cards with envi-

ronmental descriptions, such as warm and feeling oriented; control by physical threats or punishment; emphasis on status, power, and material goods. As in the measure of psychological adjustment, those descriptions that were most salient for a subject were put in the first few piles and those least characteristic in Piles 8 or 9.

Actuarial Data   Finally, there was information on the subject's IQ, the family's socioeconomic status, the number of marriages of each parent, etc.

Results

We will concern ourselves only with the major finding, so as not to get lost in details. As the writers state (Siegelman et al., 1970, p. 287), "In general, for both males and females, the families of origin of high optimal adjustment S's were more democratic—more open and direct—with greater sexual compatibility of parents, freer interchange of problems and feelings (e.g., fathers of high optimal adjustment males more direct even in anger), greater agreement on values and important life areas, and greater orientation toward intangibles and the higher cultural values. Mothers, especially, reflected this greater openness, as well as greater intellectual acuity and greater satisfaction with their maternal role. The families of low-adjustment S's can be contrasted on all these variables; in general, their homes showed signs of conflict and discord, and the mothers were more dissatisfied with their role."

We see in Table 9.2 the basis of these statements linking later adjustment to parental practices while the child was being reared. In "Early Family Ratings," we see well-adjusted parents with marked "togetherness" for the high-adjustment subjects. For the low-adjustment subjects, there is the awful pairing of restless mother and sullen father. In "Ratings of Mothers," there emerges a bright, "pulled-together" type for the high-adjustment group, with a disorganized mother who rattles on for the low-adjustment group. Environmental Reminiscences of the high-adjustment group gush with warmth, concern, and ethical values that would not be out of place in an old Andy Hardy movie. For the low-adjustment group, the reminiscences are more like those of the self-pitying antihero of today's films, replete with authoritarianism and materialistic values. (The seductive mother does come as a shock.) Finally, in "Actuarial Data" we see the high-adjustment subjects tend to be brighter and better educated.

TABLE 9.2
Background Characteristics
of the High- and Low-Adjustment Groups

High-Adjustment Group	Low-Adjustment Group
Early Family Ratings	
Good sexual adjustment	Mother restless
Parents have similar recreational interests	Mother worrisome
Mother free in volunteering sex information	Father withdrawn and sullen in face of conflict
Father responds openly to conflict	
Ratings of Mothers	
Rapid and accurate thinking	Talkative in a disorganized way
Intelligent, mentally alert	
Cooperative, frank, and trustful	
Tends to criticize child	
Poised and satisfied with lot	
Environmental Reminiscences	
Home and mother: warm, feeling oriented, love and tenderness	Home and mother: cheerless, discordant
Ethical responsibility emphasized	Parents induce conflict
Mother enjoyed the maternal role	Status, power, and material goods emphasized
	Mother neurotic and anxious
	Mother authoritarian and rejecting
	Parents discourage independence
	Mother seductive
Actuarial Data	
More education	Lower IQ

Associative Confounding with Other Variables

Let us first assume that the correlations of present adjustment with antecedent variables were correctly described. How sure are we that this kind of relation would have been found had we been able to do a valid manipulative experiment? If good methods of sampling had been employed in the experiment, the groups would have been equated on socioeconomic status, intelligence, and even psychological adjustment of parents. Remember, we would have to have found a way to convince one group of parents to use what are thought to be good child-rearing practices and the other group to use faulty practices. We would then be able to say that it was the level of child-rearing practices that brought about differences in the dependent variable, psychological adjustment. Because the two groups were not equated on these other variables, we have quite a number of other possible conclusions.

One conclusion would agree that child-rearing practices were, in fact, better for the high-adjustment group. However, it would state that another variable— say, intelligence or socioeconomic status of parents— *independently* determined both child-rearing practices and future adjustment of children. For example, to the extent that intelligence is inherited, children of more intelligent parents tend to be more intelligent. This is a positive factor in psychological adjustment. The more intelligent parent would also have higher values and better child-rearing practices, but that would not be the determinant of future adjustment.

Another possibility is that there really was no difference in child-rearing practices of high- and low-adjustment groups at all—that the difference was in the eye of the psychologist. Intelligent people are skilled at making a favorable impression. This is also true of people with high socioeconomic status. They seem "so nice." Thus, if parents' intelligence or socioeconomic status determined future psychological adjustment, the apparent harmony observed in the home could well have to do with skillful cover-up of tensions.

The "Environmental Reminiscences"—which provided the entire basis for conclusions relating to democracy in the home—introduce new possibilities of confounding. This part of the study makes us aware of the special difficulties of studies that rely on memory. The first confounding variable is *experimenter bias*. Remember that the two groups were selected on the basis of interviews with psychologists. Now, another psychologist is interviewing the same subjects. This second psychologist should also be able to evaluate the adjustment of a given subject. It would not be surprising if that evaluation affected the psychologist's ratings of home environment. He would be biased toward picturing a favorable environment for a well-adjusted subject and an unfavorable environment for a poorly adjusted subject. The ratings of home environment are anything but "blind," as described in Chapter 5. The second confounding variable is—surprisingly enough—*subject bias*. A well-adjusted person might be expected to have favorable reminiscences of his childhood, while a poorly adjusted person would have unfavorable reminiscences. One of the main aspects of psychological adjustment is the way you evaluate the world around you: past, present, and future.

A final possibility of an invalid interpretation of results has to do with external validity rather than internal validity. This would be that the "dependent variable" is a poor *operational* representation of adjustment. Suppose that a child picks up the skill of making a good impression from its parents. Now, just as the parents were able to "con" one generation of psychologists, their children are able to "con" a new generation of psychologists. All aspects of validity would be strengthened by less reliance on subjective evaluations and more use of objective evidence of adjustment as provided by the activities of daily life.

As far as internal validity is concerned, the threat in this study has been that of confounding with other variables. There were no controls for any of the confoundings. We shall now go on to consider a correlational study in which the investigators were well aware of other possible determinants of behavior and were in a good position for employing controls.

AN INVESTIGATION ON BIRTH ORDER AND INTELLIGENCE

Only on the rarest of occasions is an entire population available for studying antecedents of a behavioral difference. Such an opportunity was seized by Lillian Belmont and Francis A. Marolla (1973) of the New York State Department of Mental Hygiene. The circumstance was the availability of records of behavior on an intelligence test of almost 400,000 young men in the Netherlands along with accompanying information on the individual's position in the family birth order, the size of family, and the occupational class of the father. It all came about because of a kind of coincidence. The young men were born during the famine years of 1944 to 1947 brought about by World War II. A massive study was done to find how this dire beginning affected their future health and physical development. Happily, there was no adverse effect. Somewhat less happily all young men in the Netherlands are required to be available for service in the armed forces when they reach the age of 19, and are tested on intelligence (as well as on specific aptitudes) at that time.

Sir Francis Galton, cousin of Charles Darwin and a brilliant biologist in his own right, noted long ago that there were a disproportionate number of first-born children who achieved eminence in science ([1874] 1970). Similar observations were subsequently made, covering many fields of endeavor. In more recent years, Altus (1965) found that there was a regular birth order effect on verbal scholastic aptitude scores of young adults already selected for academic prowess. This investigation, while on sound statistical ground using a sample of some 1,878 students, invited even more thorough study. First of all, other investigators have not obtained the same result. Second, the relations were established for only a small segment of the population. Third, there is a possibility of confounding by whatever variables affect family size. One of these variables is socioeconomic class; those lowest in this respect tend to have the largest families. Thus there are progressively more children from poor families as you move from the second-born to the third, the fourth, etc. It has been well established that there are indeed differences in intelligence favoring high socioeconomic status. It is thus necessary to conduct a study to show that the birth order effect is not due to size of family. Further, it would be interesting to learn whether size-of-family effect is something more than a reflection of socioeconomic status. Surprisingly enough, this bias did not appear in Altus' study: First-born sons in families of four children tested slightly higher than did those in families of two or three children.

The Measures Used

Family size and birth order for each individual were simple matters of public record. It should be noted that use of such records is allowable only with absolute guarantees that the identity of particular persons is protected. Social class of father was described by a listing of occupations, using the three broad categories: "nonmanual (professionals and white collar workers); manual (skilled, semiskilled and unskilled workers); and farm (farmers and farm laborers)" (Belmont and Marolla, 1973, p. 1097).

Intelligence-test scores were obtained when the young men turned 19. The one used in this investigation is a Dutch version of the Raven Progressive Matrices (Raven, 1947). A sample of the kind of item used is shown in Figure 9.1. Scores were grouped into six intervals or levels, the best being placed in Class 1 and the poorest in Class 6. Thus each person was given an intelligence-test score of 1, 2, 3, 4, 5, or 6.

The Group Difference Found

We may see the entire pattern of differences in the three graphs of Figure 9.2. The mean score over a group is plotted on the vertical axis.

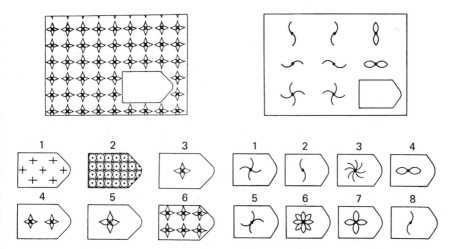

Figure 9.1 Sample items from Progressive Matrices Test (Raven, 1947). The task is to select the correct insert to fill in the matrix. From Anastasi, A. *Psychological Testing.* New York: Macmillan, 1976 (p. 292). Copyright, Anne Anastasi, 1968. Reproduced by permission from the publisher and by the publishers of Raven, J. C. *Progressive Matrices.* London: H. K. Lewis. 1947.

On the left, in (a), both the effect of family size and social class are clearly demonstrated. Raven intelligence score generally decreases as family size increases (although there is somewhat of a reversal for a family size of 1). Also, scores are generally higher with higher level of social class. In the middle graph, (b), it is seen in addition that there is a regular birth order effect with lower average intelligence all the way down to Order 9 (which also included even larger families). Thus far, the effect of family size and birth order have been found with the confounding variable social class *held constant*, i.e., controlled statistically. Also, the effect of social class has been shown with family size and birth order held constant.

Until we examine the somewhat more complicated set of curves in the graph on the right, (c), we have not yet gotten rid of the association of family size with birth order. As has been stated, this is because larger families are increasingly represented in the later birth orders. However, the graph on the right shows that birth order operates even with family size controlled. Each line that represents one family size (F.S.) is seen to slope down to the right with higher birth order. This set of curves also shows that the family size effect does not come about by having more representatives with later birth order. While this might be the case in

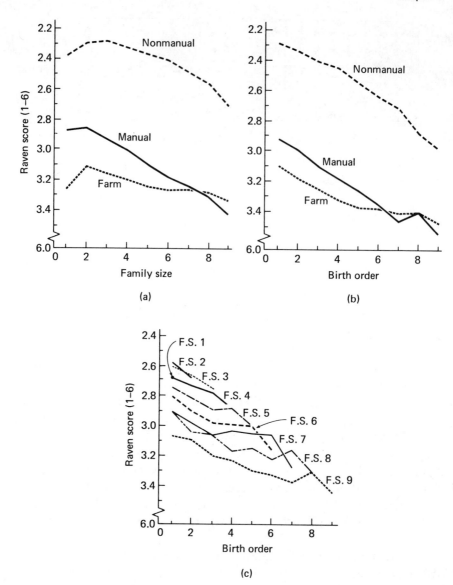

Figure 9.2 Differences found in study on birth order. From Belmont and Marolla (1973, pp. 1097 and 1100). Copyright 1973 by the American Association for the Advancement of Science. Reproduced by permission.

comparing the curves for Family Sizes 2 and 3, where Position 3 lies along a continuation of the curve for family size 2, it is not the case for the larger families. The whole curves for Family Sizes 4, 5, 6, 7, 8, and 9 lie progressively lower.

Interpretation

Let us confine our exploration of the meaning of the results to the relation between birth order and intelligence. The relation of intelligence to social class and size of family gets into the familiar heredity-environment controversy, with not especially good data (say, as compared with a study of identical twins). Belmont and Marolla (1973) do not go into any real attempt to interpret the relation of birth order to intelligence. Their task was one of statistical substantiation, with good controls. There cannot be any systematic difference, on the average, in the traits or abilities of children on a *hereditary* basis when they belong to the same family. Two basic kinds of explanation seem possible. One is that birth order makes a difference in the kinds of experiences the growing child has. The other is that the conditions in the fetal state are more favorable for early-born than for later-born infants. Suggestions of the first type (postbirth environment) are that parents pay a good deal of attention to their first-born child and then progressively less to the additional children, that later children learn their speech from other children rather than adults, and/or that younger children in a family are bullied or otherwise "kept down" by older children. The investigators suggest (for a prebirth explanation) that "mothers might become less effective reproducers with an increasing number of children" (p. 1101).

Still another interpretation that could be made has both prebirth and postbirth possibilities. It is that children further along in the birth order have a higher probability of being unwanted. This might influence the mother in prenatal care; for example, in respect to nutrition and avoidance of tobacco and alcohol. It might also affect the love and attention given to the child.

How Effective Were the Controls?

The main purpose of the investigation was to test the hypothesis concerning birth order and intelligence. The other variables—which also affect intelligence—namely, family size, and socioeconomic status, were held constant by means of homogeneous groups. We have already seen that there is more than one interpretation possible of the birth order effect, if it can be assumed that there was no confounding with other variables. Now let us consider whether that assumption is justified.

Some Effective Controls   The investigators succeeded in controlling two additional variables that might have been overlooked even in the world-of-the-future manipulative experiment. Suppose all of the couples are enlisted on their wedding day in 1983. Then, some years later, comparisons will be made between their strictly scheduled first-born, second-born, third-born, etc. children. If this is done, there will still be confounding with other variables.

Let us suppose that the children were born in 1985, 1987, 1989, etc. If they were all tested at once, say in the year 2010, the first-borns would be 25 years of age, the second-borns 23, the third-borns 21, etc. This age difference confounding, of course, could be avoided by testing each subject when he reached the same age—say, 19, as in the Dutch study. Still, the different groups of subjects would belong to slightly different *cohorts* (people born at about the same time). Perhaps school curricula would change while the children were growing up. The educational background of the 1985 cohort would be different from that of the 1989 cohort. In the Belmont and Marolla study (1973), all subjects were tested at the same age, so that variable was controlled. They were all from very close cohorts, being born within a span of three years. Also, there could not have been any systematic relation between birth order and cohort. A ninth child in a family had the same probability of being born in 1944 as in 1947.

Failures of Control   We have seen the difficulty of controlling all possible other variables that might influence behavior in the study on optimal adjustment. Here too, despite the investigators' sensitivity to possibilities of confounding, there remain variables that have not been controlled. One possibility that does not seem to have been considered is the age of the parents. It could be argued (for a postbirth explanation) that younger members of the family tend to have older parents who may have lost some of their playfulness. A prebirth explanation is that a mother may be a "less effective reproducer" simply because she becomes older, not because she has had previous children.

Another confounding variable, not controlled for, is the number of years the parents have been married. Certainly the emotional atmosphere of the home changes over the years of marriage. Children farthest along in the birth order are also farthest from the honeymoon.

A final note of a confounding of variables in the study is that the control over social class is not all that it might have been. It would seem that there are as large cultural differences within each of the three social classes as there are between classes. Thus, social class was not controlled (or more probably is not controllable) in the same way as is size of family, age of parents, or number of years of marriage.

Many of you must be struck by the fact that the subjects were all males. This, of course, was not due to male chauvinism on the part of the investigators. The data were available only for males. Still, we cannot conclude that the birth order effects would hold for women. The external validity is thus limited.

CONTROLS IN CORRELATIONAL STUDIES

Now, using these two investigations as a background, let us consider controls in correlational studies in a more general way. The problem is the same one that we have faced in manipulative experiments, that of systematic confounding with another variable. However, in a correlational study, we cannot use *manipulative* controls for confounding but must instead use *statistical* controls. We do not start with groups equal on other variables; we hope to equalize groups on these variables. Equalizing is done in one of two ways in correlational studies.

Matched Individuals

When the number of subjects is relatively small, the method of control is often that of matching individuals. In the investigation on psychological adjustment, this could have taken the form of finding pairs of high- and low-adjustment subjects who were very similar on intelligence of parents and on their socioeconomic status. Suppose that 50 such pairs were found. Then there would be no overall difference between groups on these two other variables. There are two problems with this technique.

First, if the other variable is important, there will have been a disproportionate number of those high and low in it in the two groups. Thus, most of the original high-adjustment subjects might have come from families with high socioeconomic status. This will make it difficult to find many matched pairs. Thus, among subjects with high socioeconomic status there will not be many low-adjustment subjects found to make pairs with high-adjustment subjects. This will be especially serious if we try to match individuals on several other variables, rather than on just one other variable. We will quickly reduce the number of pairs. On theoretical grounds, it can never be known whether every possible other variable of importance has been controlled. Here, on practical grounds, we find ourselves severely limited in the number of variables that can actually be controlled.

Second, and related to this first problem, is the unrepresentative nature of the subjects who remain. Let us say that it is typical of a high-adjustment person to have had intelligent parents, with high socioeconomic status, and with good psychological adjustment of their own. Most of these high-adjustment subjects must be eliminated in order to make the pairings. These might be the very subjects whose parents did use good child-rearing practices. Thus, the equating process might "wash out" the effect of child-rearing practices.

Thus, with the method of matching individuals, there are two dangers: that we have not exercised enough control or that we have exercised too much control. Too little control means that an important confounding variable has not been identified; too much control means that the effective variable has not been allowed its full range of variation.

Homogeneous Subgroups

Since the birth order investigation had so many subjects available, there was no need to try individual-by-individual matching. Rather, *homogeneous subgroups* were set up who differed on level of the variable of interest but who were equated on another variable. Thus, difference between second and fifth child could be compared for the homogeneous subgroups of 5-child, 6-child, 7-child, etc. families. If this had not been done and all second-born children had been compared with all fifth-born children, there would have been a confounding with family size. Fifth-borns can come only from families of five children or more; second-borns can come from smaller families.

The possibly important other variable that was missed, age of mother, could also have been used as a basis of subgroups. This would be best employed after the original breakdown into families of different size. Thus, for example, the first- through last-born children of Family Size 5, could be compared for mothers who were 23 years of age at the time the subjects were born. Similarly, a comparison could be made for 24-year-old mothers, etc. If the birth order effect disappeared with age of mother held constant, we would realize that there is an age of mother effect, not a birth order effect. More likely, even if age of mother was shown to make a separate contribution, birth order differences would still occur. The variable years of marriage could also be controlled by homogeneous subgroups. Homogeneous subgroups could be set up for one year, two years, etc. of marriage at the time the subject was born.

If there are different effects of the variable of interest for the different homogeneous subgroups, this fact may lead to deeper understanding of how the variable operates. Let us take the overall birth order effect as genuine, not due to age of mother or to years of marriage. We see in Figure 9.2(b) an actual example of different effects. For subgroup "nonmanual," there is a rather steep decline in intelligence-test score with birth order. This is also true for the subgroup "manual." However, for the subgroup "farm," the decline is rather shallow.

This is to say, of course, that there is an interaction between class and birth order in their relation with intelligence test scores. We have already seen examples of interaction in the preceding chapter. It is not surprising that they would occur in correlational studies as well as in manipulative experiments. We can, in fact, use the same kind of table by which to compute the main effects and interactions. Such a presentation is found in Table 9.3 for just two levels of class (homogeneous subgroups) and two levels of birth order, 1 and 9. The entries in the table are the mean values of the intelligence test scores for the subgroup. You will recall that the best score a person could have was 1 and the poorest was 6. The main effect "birth order" is seen to have a value of .59, and the main effect "class" is seen to have a value of .67. The interaction, "birth order × class" has a value of .26, which is quite appreciable.

TABLE 9.3
Main Effects of Birth Order and Class
on Intelligence Test Scores and their Interaction

| Class | Birth Order | | | Main Effect |
	1	9	Mean	Class
Nonmanual	2.28	3.00	2.64	.67
Farm	3.08	3.54	3.31	
Mean	2.68	3.27	2.975	
Main Effect Birth order		.59		

Interaction
Birth order × class: $(3.00 - 2.28) - (3.54 - 3.08)$
$= .72 - .46 = .26$

Perhaps if "farm" were restricted to actual workers in the fields, excluding owners of large mechanized farms and land managers, the interaction would be even larger. Perhaps there would be no drop in intelligence test score with birth order. There is thus a suggestion that the birth order effect is not universal but is rather characteristic of city families or, more generally, nonland families. Even with the available data, the interaction should be taken into account in any interpretation of the birth order effect. This kind of information has become available because the method of control was that of homogeneous subgroups. The method of matching individuals gives no knowledge of interactions.

A possible explanation of the interaction is that a new family member is "wanted" more by a nonmechanized small-farm family than by other families. On the small farm, each new child represents an economic asset. Elsewhere, a new child means a new responsibility and an increase in crowding. Thus, an eighth child for a farm family is likely to be wanted; for an urban family, that is not very likely. Perhaps the cumulative effect of being wanted or not wanted as a small child translates into intelligence test scores at age 19.

In theory, it would have been possible to have obtained a statement from parents as to whether or not the child was wanted. Suppose this had been done. Now we would be able to find within each family size homogeneous subgroups of wanted or unwanted children for all birth order positions. Suppose, further, that there was no difference between second-born and fifth-born children in Family Size 5 if they fell into the same subgroup of wanted or unwanted. It would be possible to conclude that the birth order effect reduces simply to whether a child was wanted. However, this would not be a *necessary* conclusion. Remember that all we have in a correlational study are correlations, and this is one of them. Possibly life on the farm is less wearing than city life. Thus, the city person looks with dismay at the approach of the tiny stranger because of the effort that will be required and for that reason does not want the new child. The healthy farm person feels no such apprehension. A late child in the city will suffer because of lack of energy by the parents, and this will not happen on the farm. The wanted-unwanted difference would then not be the basis of birth order effect; rather, it would be the amount of energy of the parents—which was *correlated* with the child's being wanted or unwanted.

A great deal of control, then, is made possible for correlational studies by the method of homogeneous subgroups. Even so, we can never know whether an antecedent variable really does affect a measure of behavior. As in the case of matched individuals, we have no way of knowing whether all important other variables have been taken into account. Moreover, when we have finally narrowed down an observed effect to what seems the critical variable in its operation, there still remains the possibility that some other correlated variable was the real determinant. This kind of difficulty has given rise to the familiar statement that correlation should not be mistaken for causation.

A STUDY TO PREDICT PROFICIENCY IN INSPECTION

Let us consider a made-up example of how a correlational study could be used to improve practical predictions. There is a factory that is having trouble in maintaining quality control over intricate mechanical assemblies. The majority of inspectors have been accepting assemblies that are faulty. When they are told to be more careful, they become unsure of themselves and begin to reject assemblies that later testing shows are satisfactory. These inspectors seem intelligent and well motivated but seem to lack some particular ability. The solution is not as simple as trying a number of people on this job and keeping only those who do satisfactory work. First, it is just too uneconomical in view of the large number who prove unsatisfactory. Second, an employee stands to lose valuable experience that could be gained on some other job. The situation would be tolerable for all concerned if 80 percent of those assigned to the job of inspector would turn out to be satisfactory.

The problem is given to the personnel director who has had experience with aptitude tests. He decides that there is a commercially available test that should be tried out. In it there are drawings of connected parts shown in various orientations. In each set, there is one drawing in which some angle between pieces or point of attachment is different from that in a standard drawing. The task is to mark the error in the faulty drawing. Scores can range from 0 to 85 points. As a matter of fact, few people get under 40 or over 80 points.

Method

People applying for all kinds of jobs are told in advance that they have a chance at the desirable job of inspector. There are other jobs available if they do not prove satisfactory. In all, 60 new employees take the aptitude test. They then work half-time on the inspection job and half-time on some other job. After they have been doing this for 3 months, they are all evaluated on their inspections in the fourth month. During that time, a record is taken of the number of items they inspect and also of their error rate. There are two kinds of error. First, there is failure to detect a flaw on any of 40 "keyed" parts that are known to be defective. Second, there is finding a flaw in another 40 parts that are known to be perfect. They all know that this is to be done. A person's *criterion score* is the number of parts inspected in the last 20 days minus four times that number multiplied by the person's error rate. Thus, the person who inspected 800 parts with an error rate of 5 percent obtains a criterion score of 640, thus: $800 - .05(3200)$. A criterion score of 675, or over, is considered to be satisfactory.

Results

The scores for each individual on the *aptitude test* and on the *job criterion* are plotted in the form of a scatter diagram. This is shown in Figure 9.3(a). Each number in a square represents the number of individuals with a particular test score and criterion score combination. Its horizontal position shows how high those persons' test scores were; its vertical position shows the height of their criterion scores. Thus, the circled "1" represents the one person whose test score was between 75 and 79 and whose criterion score was between 750 and 774.

In general, all scores are shown to fit inside an oval that rises from left to right. This means that the scores are positively correlated. The amount of correlation may be computed. The method is described in the statistical supplement at the end of this chapter. For the scatter diagram in (a), the value of the coefficient of correlation, the number describing the amount of correlation between test and criterion is .60, or +.60, for those who are fussy. We really don't have to consider the possibility of *negative* correlations very seriously in a practical problem. There would be a correlation of $-.60$ in (a) if the scatter diagram generally went down from left to right. Negative correlations, when found, are often low, in which case they may be suspected of being accidental fluctuations from a "true" correlation of zero. When they are sizable, they most typically come about because of the direction of scaling on one of the axes. Thus, number correct on one test will have a negative correlation with number of errors on another.

As can be seen, a correlation of .60 indicates pretty good agreement between test scores and criterion scores, but far from perfect agreement. Some individuals who are high on the test prove to be poorer inspectors than others who score lower. In (b) is shown *perfect* agreement between test and criterion scores (which never happens). If one person is a certain amount higher than another on the test, he will be just that amount higher on the criterion—if scores are scaled off in a standard way. When the coefficient of correlation is computed for (b), the value is found to be $+1$.

Still the correlation of .60 shows better agreement than that represented in (c). In this latter case, there is some trend for high criterion

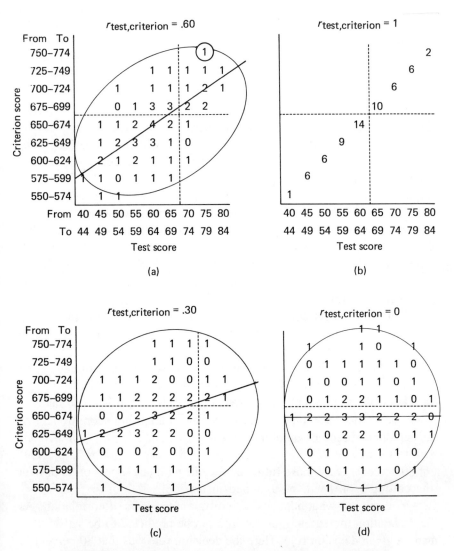

Figure 9.3 Scatter diagrams of test scores and criterion scores of job performance in inspecting. Criterion scores above the horizontal line represent satisfactory work. Test scores to the right of the vertical line drawn in (a), (b), and (c) are predictive of satisfactory job performance. No prediction is possible in (d).

scores to go with high test scores, but not much. When this coefficient is computed, its value is found to be +.30. Even that low amount of correlation is higher than that represented in (d), where the coefficient of correlation has a value of 0. Here there is no trend at all.

Assigning Future Applicants   Going back to (a), we can see how the results of the study could be used by the personnel director to screen applicants for assignment to the job of inspector. The horizontal line separates satisfactory performance from unsatisfactory performance. Remember that criterion scores of 675 or over represent satisfactory performance. If we count the number of persons over this line, we find that there are 24 of the 60, only 40 percent. A vertical line has been drawn just to the left of the test score of 70, which is called a *cutting score*. There are 14 persons to the right of that line. Of those, 12 were satisfactory. This gives 86 percent, somewhat better than the 80 percent required. However, if the cutting score were moved one space farther to the left (down to 65), it would include 24 persons of whom 17 were satisfactory, bringing those satisfactory down to only 71 percent, far less than the 80 percent required. Thus, the personnel director can *infer* that only future applicants scoring 70 or above should be assigned to the job of inspector. The assumption is that the relation found with the present subjects will hold for future applicants.

With the perfect correlation shown in (b), there is no point in even discussing the percentage of satisfactory persons of those selected. There is no way of making an error. All of the 24 persons with a test score of 65 or over were satisfactory. In the future, the personnel director can assign anyone with a score of 65 or over to the job with zero risk.

However in (c), with the correlation of only .30, it is seen that even when the vertical acceptance line is moved to a cutting score of 75 that the personnel director cannot meet the required goal. Of the eight persons with test scores of 75 or over, six were satisfactory, which is 75 percent, not 80 percent. Only those future applicants with test scores of 80 or over (the cutting score) will be assigned to the job.

You can see what happens with different amounts of correlation. As the correlation increases, more people can be predicted to be satisfactory using a given decision rule. Here the decision rule was that 80 percent of those selected should be satisfactory. In practice, the selection would not be made by shifting lines on the obtained scatter diagram, which undoubtedly has many accidental irregularities. Instead, predictive tables would be used. They are based on idealized scatter diagrams to represent different correlations between a test and a criterion (Taylor and Russell, 1939).

Predicting Individual Performance   Now let us look at the situation from the applicant's point of view. The applicant would like to know the level of job performance he is most likely to obtain. (He should also want to know by how much that prediction could be in error.) The slanted line in (a) is drawn to connect the average criterion scores for different test scores. For example, the mean criterion score for people making between 55 and 59 points on the test falls in the interval 625 to 649. The mean criterion score for those between 70 and 74 on the test falls in the interval 675 to 699. The connecting line between the mean criterion scores for these test scores, as well as for all of the other test scores may be used as a *prediction line*. Any future applicant, after being scored on the aptitude test, could look at the scatter diagram with its slanted line and make a prediction of how well he would do at the job of inspector. He would notice that there was some error in the prediction. Thus, people with test scores of 70 to 74 obtained criterion scores as high as 725 to 749 and as low as 600 to 624, even though the predicted score was between 675 and 699.

With the perfect correlation in (b), there is no error in prediction. Each test score gives an exact prediction of criterion performance. However, with the low .30 correlation in (c), people with test scores of 70 to 74 obtained criterion scores of as high as 750 to 774 and as low as 550 to 574 (with a prediction of just about exactly 675). This is more error of prediction than was the case with the higher correlation in (a). With the zero correlation shown in (d), a criterion score of about 660 is predicted regardless of the person's test score. The oval of (a) has become a circle, meaning that there is now a very large error of prediction.

You may also have noted that the higher the correlation, the steeper the slope of the prediction line. The range is from the horizontal line for the zero correlation in (d) to the one-to-one slope (45°) for the perfect prediction in (b). A zero correlation means a zero slope; a correlation of one means a one-to-one slope, provided the test scores and criterion scores are scaled in standard equal units. Thus, the higher the correlation, the greater the range of predicted criterion scores permitted by the prediction line. Again, the irregular scatter diagram actually obtained would not be used for predicting criterion scores or for indicating amount of predictive error. A table would be prepared based on an idealized scatter diagram for the value of the coefficient of correlation. In fact this prediction may be made from a simple formula (Formula 9.3), as shown in the statistical supplement to this chapter.

What Is Accomplished in a Predictive Study?

By using the coefficient of correlation, some exact predictions are made possible. We are able to infer from the correlation between a test and criterion the cutting score that will be required to provide a given percentage of satisfactory placements. We are also able to infer the most likely criterion score for each test score for future applicants and the amount of predictive error. In the example given, one behavior was used to predict another behavior. Perhaps the reason for the correlation was, as conjectured, that the aptitude measured is important for the job. However, there are many other possibilities. Perhaps it is some other characteristic, such as attentiveness; perhaps it is due to the amount of effort a person is willing to expend. In a practical case, we are not really concerned with the explanation. Results are what count. With a high correlation, there is a good prediction; with a low correlation, there is poor prediction.

Reliability and Validity of Tests

The terms *reliability* and *validity* have been applied to tests but in not exactly the same way as these terms have been used elsewhere in this book. A test is called *reliable* if we can depend on the same person to make about the same score (relative to others) on it time and again. As we know, there are many reasons for inconsistency of behavior, including changes with time that cannot be controlled. However, this inconsistency is reduced by having a sufficiently long test, of course at a suitable level of difficulty. One way of finding test reliability is by giving the same test (or closely similar versions, if necessary) to the same group on two occasions. If the test-retest coefficient of correlation is high (e.g., .90), the test will be called reliable. However, there is also the question of whether the *study* itself is a reliable one. This means that a large number of subjects should be tested. That is *test* reliability should be based on good *study* reliability.

A test is said to be *valid*—always in respect to some particular criterion, such as proficiency in inspection—if it correlates highly with the criterion (e.g., .60). Again, a sufficient number of subjects should be used to know whether this is a valid conclusion, i.e., coming from a reliable study. We might also reach a valid conclusion that a test is invalid, i.e., a poor predictor of the criterion.

KINDS OF CORRELATIONAL STUDY

We have now discussed three different correlational studies. This is far from a complete coverage. Some idea of the wide range of correlational studies may be grasped by considering the dimensions in which they differ. However, let us first reconsider what different correlational studies have in common, or rather what they lack in common. That is *planned manipulation* of an independent variable. The investigators who studied child-rearing practices did not convince some parents to use practices considered better and others to use those considered worse. These differences in practices just *existed*. Birth order of a child was not determined by an experimenter. It, too, just existed. In the same way, existing differences on an aptitude test were not brought about by experimental manipulation. A correlational study, then, is one in which differences in a behavior are related with existing differences in another variable, sometimes another behavior. Now let us look at the dimensions along which they differ.

Approximation of an Independent Variable

As has been stated, *in principle* a manipulative experiment could be designed in which the experimenter decides on which parents will use good child-rearing practices and which ones will use poor ones. A correlational study was done instead because the manipulative experiment was impractical. Birth order presents more of a problem. By what manipulations can it be determined that a particular child shall be a fourth-born? Still, birth order is something like an independent variable. It is an antecedent variable and so a possible determinant of differences in behavior. This is not at all the case for the aptitude test used to predict proficiency in inspection. Whatever it was that determined that a person will obtain a high score on the test is the same thing that determined that he will be a good inspector. There is nothing of the independent variable in differences in aptitude test scores. As a matter of fact, we could just as well find out how well a person will perform on the aptitude test from his performance as an inspector. It is not done only because it is not a useful thing to do.

Description of the Correlation

The three investigations were all called *correlational*, but only in the inspection study was a coefficient of correlation computed. This measure is most meaningful when each of the two variables related forms a continuous bell-shaped distribution. This is true of scores on almost any test. They differ continuously from low to high, and they tend to pile up around the average value. Thus, in the study on inspectors, the coefficient of correlation was appropriate for describing the correlation between the two variables. This approach could have been used also in the study on adjustment. Each subject had an adjustment score on a nearly continuous scale. It is almost certain that a bell-shaped distribution was approximated. The antecedent variables had somewhat the same kind of distribution, although generally in steps rather than in fine gradation. Family income could have been used directly to provide continuous scores, although the distribution tails off to the high-income end. High- and low-group comparisons were probably used instead of coefficients of correlation, because it is not very clear what is meant by middle amounts of adjustment.

The curves of Figure 9.2 are about all that can be done to describe the correlations of family size and birth order with intelligence test scores. The latter, of course, do make up a continuous bell-shaped distribution. However, this was not the case of these two antecedent variables. Use of a coefficient of correlation would not be very meaningful for this reason.

Goal

The adjustment and birth order studies were done to obtain understanding of the determinants of differences in behavior. That is not to say that the findings from the adjustment study could not be put to practical use. It is difficult to see any immediate application of the results of the birth order study. (This writer already has had the bad luck of being the youngest child in his family!) Of course, the study to predict job performance clearly had a practical goal.

Let us not get these dimensions "confounded." It is not implied that studies in which a coefficient of correlation is computed or one in which two measures of behavior are related are bound to be practical in nature. Many studies of theoretical rather than practical importance share these characteristics. Coefficients of correlations have been found between test scores of parents and children, between scores for pairs of identical twins, etc. These are theoretical studies that try to separate hereditary and environmental influences on behavior. Also, theoretical studies are performed in which different tests are given to the same group of individuals —just as was done in the inspection study. However, as many as 40 or 50 different tests are sometimes given, and a coefficient of correlation is computed between each pair of tests. A technique called *factor analysis* is used to discover a much smaller number of underlying variables adequate to describe differences among persons.

SUMMARY

Correlational studies are used to test hypotheses concerning behavior in cases where manipulative experiments cannot be performed. In two examples, the investigations on antecedents of optimal adjustment and on the effect of birth order on intelligence, manipulative experiments were impractical. In the study in which a correlation was obtained between aptitude test scores and job criterion scores, it was hard even to imagine the experimental manipulations that would be necessary.

When high- and low-adjustment subjects were compared on a number of antecedent variables, the main finding was that there had been superior child-rearing practices for the well-adjusted subjects. However, there were other variables confounded with child-rearing practices. Two of them were socioeconomic status and intelligence of parents. It cannot be told whether they actually biased the results as no controls were employed. Further possibilities of confounding were brought about by the use of subjective ratings. Because the ratings were not "blind," there was the possibility of experimenter bias. Because the ratings were based on interviews with subjects, there was also the possibility of subject bias— more favorable reminiscences by well-adjusted persons.

Intelligence test scores were found to decline with birth order in a study on almost 400,000 young men. This effect held up even with control of occupational class and family size. Generally, the controls in this study were impressive. However, it was noted that occupational class consisted of some rather loose groupings, probably overlapping on important other variables. Various interpretations may be made of the correlation of intelligence test score with birth order, divided mainly into prebirth and postbirth factors. Two variables were not controlled for. The first is age of parents, where that of the mother may be of particular importance. The other is that of length of marriage at the time of the subject's birth.

Controls for correlational studies may be broken into two classes. First, there is individual matching of subjects. If two groups are compared, each individual in one group is matched with another individual in the other group who is at a similar level on certain other variables. There are two difficulties with this technique. The first is that not many matching variables can be used without losing many of the subjects. Thus it is likely that there will not be control over enough other variables.

The other is that the subjects who remain will not be typical of the two groups so the relations found will not be representative. Thus there are the dangers of both too little and too much control.

The other method is that of homogeneous subgroups, as employed in the study on birth order. For example, a subgroup of subjects in Family Size 5 was set up. Within this subgroup, it was found whether the birth order effect held up. The different subgroups amount to different levels of the other variable. Thus it is possible, as with two independent variables in a manipulative experiment, to find whether there are important interactions between contributing variables. In correlational studies, this may also help in understanding the relation between variables. For example, the small amount of decline of intelligence test scores found for farm workers might be due to a greater "wanting" of an additional child. If data on wanting were available, the possibility might be followed up that this is the basis of the birth order effect. This would be shown by a decline in intelligence test score with birth order for homogeneous nonwanting subgroups within both the city group and the farm group, but with no decline for wanting subgroups.

However, because the evidence is still correlational, this would not establish that wanting was proved to be the basis of the effect. For example, wanting may simply be correlated with physical energy of the parent, and that may be the basis of the effect. So it goes with correlational studies. Interesting hypotheses may be investigated. Many ideas may be suggested. Still, control can never be as good as that in a manipulative experiment.

The final study described was a made-up example of how an aptitude test might be used to find which applicants will be able to perform the job of inspector. A group of subjects were tested on aptitude before being put on the job. Their job performance was also evaluated. A scatter diagram showed the relation between the two variables, aptitude test score and job criterion score. The shape of the scatter diagram can be expressed by the coefficient of correlation. The positive values range between 0 and 1. It was shown through scatter diagrams representing different values of correlation how applicants may be selected according to a decision rule, e.g., that at least 80 percent of those selected will be satisfactory. The higher the coefficient of correlation, the lower it is possible to set the cutting score on the aptitude test, which allows the selection of more individuals, and still meet the requirements of the rule. The correlation also allows prediction of the criterion score that an individual will achieve, by the prediction line relating mean criterion score to test score. The higher the coefficient of correlation, the steeper the line and the smaller the amount of predictive error.

Similarities and differences in correlational studies were described. They are all similar in relating existing variables rather than in manipulating an independent variable to find the effect on a dependent variable. They differ widely in a number of ways. First, there is the extent to which one of the variables is like an independent variable. In the study of psychological adjustment, the antecedent variables were clearly possible independent variables. At the other extreme, in the study of selection of inspectors neither variable at all resembled an independent variable. Prediction was made in one direction rather than the other only for practical reasons. Second, they differ in whether or not a coefficient of correlation is computed. This description of the amount of correlation is most meaningful when values on each variable form a continuous bell-shaped distribution. This condition is almost always met by two sets of test scores. However, such variables as income and rated personality characteristics often meet the condition well enough. Third, they differ in whether the goal of the study is that of understanding or of immediate practical application. The fact that the practical example in this chapter used the coefficient of correlation and was concerned with two measures of behavior for each subject does not mean that such studies cannot have the goal of increasing understanding.

QUESTIONS

1. Why is the study comparing high- and low-adjustment groups called *correlational*?

2. Why is associative confounding always present in a correlational study but only sometimes present in a manipulative experiment?

3. How may the ideal experiment be used as a standard of internal validity in a correlational study?

4. What are the difficulties in control of confounding by individual matching? Give an example.

5. Give an example of how the use of homogeneous subgroups can lead to information on interaction.

6. What confounding variables were missed by the investigators of the birth order effect?

7. Why is even the best correlational study limited in identifying the variables affecting behavior?

8. Why does a higher coefficient of correlation allow the identification of a higher proportion of individuals who will be satisfactory on a criterion?

9. What are the dimensions on which correlational studies differ?

STATISTICAL SUPPLEMENT

COEFFICIENT OF CORRELATION

Standard Scores

The simplest formula for computing the coefficient of correlation between two sets of scores is written in terms of *standard scores*. It is also the formula that gives the clearest idea of the meaning of the coefficient of correlation. These are two reasons for introducing you to standard scores in this supplement. Another good reason is that standard scores can be compared from one test to another. Thus, it does not tell a person very much if you say you got a test score of 38 in history and a test score of 221 in English. However, (if that person has read this supplement) it conveys a lot of meaning to say that your standard score was +2.1 in history and −1.3 in English.

You already know that a (raw) test score for a subject in a group is generally symbolized as X. Moreover, the test score for a particular subject is shown by a subscript. Thus, Subject 3's test score is symbolized as X_3. You are also familiar with the deviation score, $x = X - M_X$. Subject 3 has a deviation score of $x_3 = X_3 - M_X$. Now, if a subject's deviation score is divided by the standard deviation, σ_X, of the distribution of scores, it is transformed into a standard score (or z score).

Suppose Subject 3 has a (raw) test score of 60. The mean score for the group is 49 and the standard deviation of scores is 12. Thus, $X_3 = 60$, $M_X = 49$, $\sigma_X = 12$. First of all $x_3 = 60 - 49 = +11$. Let us now solve for z_{X_3}, that is, for Subject 3's standard score.

$$z_X = \frac{x}{\sigma_X} \qquad \text{(Formula 9.1)}$$

Here

$$z_{X3} = +\frac{11}{12} = +.92$$

Since standard scores are seldom much higher than +2 or lower than −2, you know that this subject's score is roughly halfway between the mean score and the highest score in the group.

The criterion scores, such as in job proficiency, which are correlated with test scores, are usually symbolized by Y instead of X. Thus a deviation score is symbolized by y and a standard score on the criterion by z_Y. We thus speak of finding the correlation between X and Y, each subject of the group having an X score and a Y score. The symbol for the coefficient of correlation is r_{XY}.

Computation of r_{XY}

Again, we will use data previously presented for this computation. Let us take the Treatment A entries to be test scores of 17 subjects. Now we will consider the Treatment B entries to be the criterion scores for the same subjects. However, so that the comparative nature of standard scores may be emphasized, we will multiply each of the Treatment B entries by 10. Fortunately, we have already done a good deal of the computing necessary to obtain r_{XY}. For the test scores, we will simply use the mean and standard deviations obtained. For Treatment B, we have merely to multiply our obtained mean and standard deviation by 10.

The test scores are shown in the column second to the left in the following table and the criterion scores in the second column from the right, labeled X and Y respectively.

S	X	x	z_X	$z_X z_Y$	z_Y	y	Y	S
1	223	+38	+2.054	+2.455	+1.195	+190	1810	1
2	184	−1	−.054	−.109	+2.013	+320	1940	2
3	209	+24	+1.297	+.898	+.692	+110	1730	3
4	183	−2	+.108	−.061	−.566	−90	1530	4
5	180	−5	−.270	−.102	+.377	+60	1680	5
6	168	−17	−.919	−.810	+.881	+140	1760	6
7	215	+30	+1.622	+.102	+.063	+10	1630	7
8	172	−13	−.703	+.442	−.629	−100	1520	8
9	200	+15	+.811	−.357	−.440	−70	1550	9
10	191	+6	+.324	−.143	−.440	−70	1550	10
11	197	+12	+.649	+.653	+1.006	+160	1780	11
12	188	+3	+.162	−.020	−.126	−20	1600	12
13	174	−11	−.595	−.075	+.126	+20	1640	13
14	176	−9	−.486	−.214	+.440	+70	1690	14
15	155	−30	−1.622	+.714	−.440	−70	1550	15
16	165	−20	−1.081	+2.720	−2.516	−400	1220	16
17	163	−22	−1.189	+1.346	−1.132	−180	1440	17
M	185						1620	
σ	18.5						159	

$$\Sigma z_X z_Y = +7.336$$
$$r_{XY} = +.432$$

You see that S_1's test score (X) was 223 and his criterion score was 1,810. Moving inward from both of these raw scores, we find an x of $+38$ (i.e., $223 - 185$) and a y of $+190$ (i.e., $1810 - 1620$). Again moving inward, we find a z_X of $+2.054$ (i.e., $+38$ divided by 18.5) and a z_Y of $+1.195$ (i.e., $+190$ divided by 159). Finally—in the middle column—we find the product of z_X and z_Y, which is $+2.455$.

These computations were also done for the other 16 subjects, to fill out the body of the table. Just below these entries, you see the values of the means and standard deviations. Below that, at the center, is the sum of the $z_X z_Y$ column, $+7.336$. This value, divided by the number of subjects, 17, gives the value of the coefficient of correlation, $+.432$.

So you won't have to remember all of these words, here is the formula for the coefficient of correlation between X and Y, symbolized as r_{XY}:

$$r_{XY} = \frac{\Sigma z_X z_Y}{N} \qquad \text{(Formula 9.2)}$$

Here

$$r_{XY} = \frac{+7.336}{17} = +.432$$

The Scatter Diagram

Figure 9.4 shows the scatter diagram, with a dot representing each subject. It is scaled off in z-score units. With this scaling, the slope of the prediction line is a direct indicator of the value of r_{XY}. Here r_{XY} equals $+.432$. That is the slope of the line. It goes up .432 units for each one unit it goes to the right. Thus, if a person has a z_X of $+1.000$, his predicted z_Y is $+.432$. The predicted value is thus closer to the mean of its distribution than is that from which the prediction was made. It is said that predictions *regress* toward the mean, and the prediction line is called the *regression line* in predicting Y from X. More exactly, it is predicting z_Y from z_X.

You will also notice that the prediction line runs through the intersection of $z_X = 0$ and $z_Y = 0$. Both those two points are the mean values of their respective distributions. This is true no matter what the size of r_{XY}. If a person stands at the mean in X, the best predic-

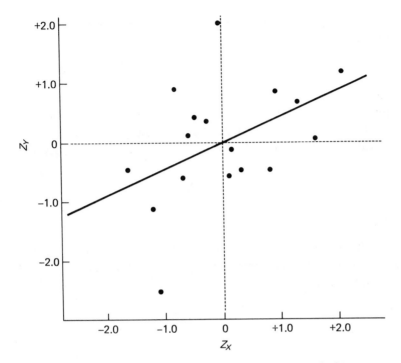

Figure 9.4 Scatter diagram with axes scaled in equal standard-score units.

tion is always the mean value in Y. You may further note that if a score is above the mean in X (a positive value of z_X), the predicted score is above the mean in Y (a positive value of z_Y). Similarly, the predicted Y score for an X score below its mean is always below the Y mean.

Finally, the higher the value of r_{XY}, the less predictions regress toward the mean. If our correlation had been perfect, the prediction line would have had a slope of $+1$. Thus, if z_X were $+1.5$, for example, predicted z_Y would also be $+1.5$. If z_Y were $-.8$, predicted z_Y would also be $-.8$. With a perfect correlation, there is no regression toward the mean. On the other hand, if the correlation had been zero, the prediction line would have zero slope. It would be absolutely horizontal. It would be at the level of $z_Y = 0$, that is the mean score for Y. Thus, whatever the value of z_X, the best prediction would always be $z_Y = 0$. With a zero correlation, then, predictions regress all the way back to the mean.

All of this may be put in quantitative terms through use of the prediction formula:

$$\hat{z}_Y = r_{XY} z_X \qquad\qquad \text{(Formula 9.3)}$$

This tells you that to predict a standard score in Y you should multiply the individual's standard score in X by the coefficient of correlation between X and Y. We may apply this formula to an individual whose standard score in X is $+.50$, with the coefficient of correlation being .70:

$$\hat{z}_Y = .70(+.50)$$

$$= +.35$$

PROBLEM: Compute r_{XY} from the data in the problem given in the statistical supplement to Chapter 6. Use Treatment C for X, and Treatment D for Y.

Answer: $r_{XY} = +.576$.

REFERENCES

ALTUS, W. D. Birth order and scholastic aptitude. *Journal of Consulting Psychology*, 1965, *29* (3), 202–205.

American Heritage Dictionary of the English Language. W. Morris (ed.). Boston: Houghton Mifflin, 1973.

AMERICAN PSYCHOLOGICAL ASSOCIATION. *Publication Manual.* (2nd. ed.) Washington, D.C.: American Psychological Association, 1976.

BAKAN, D. *On Method: Toward a Reconstruction of Psychological Investigation.* San Francisco: Jossey-Bass, 1967.

BÉKÉSY, G. v. Olfactory analogue to directional hearing. *Journal of Applied Physiology*, 1964, *19*, 369–373.

BÉKÉSY, G. v. *Sensory Inhibition.* Princeton, N.J.: Princeton University Press, 1967.

BELMONT, L., AND MAROLLA, F. A. Birth order, family size and intelligence. *Science*, 1973, *182*, 1096–1101.

BLOCK, J[ack]. *The Q-sort method in personality assessment and psychiatric research.* Springfield, Ill.: Thomas, 1961.

BRELAND, K., AND BRELAND, M. The misbehavior of organisms. *American Psychologist*, 1961, *16*, 681–684.

CALFEE, R. C., AND ANDERSON, R. Presentation rate effects in paired-associate learning. *Journal of Experimental Psychology*, 1971, *88*, 239–245.

CAMPBELL, D. T., AND STANLEY, J. C. Experimental and quasi-experimental designs for research in teaching. In N. L. Gage (ed.), *Handbook of Research on Teaching.* Chicago: Rand McNally, 1962.

CARDER, B., AND BERKOWITZ, K. Rats' preference for earned in comparison with free food. *Science*, 1970, *167*, 1273–1275.

COHAN, S. E. Packaging law is on the book, but ills it aimed to cure are still troublesome. *Advertising Age*, 1969, *40*, 10.

COHEN, J. *Statistical Power Analysis for the Behavioral Sciences.* (rev. ed.) New York: Academic Press, 1977.

COHEN, L. J. The operational definition of human attachment. *Psychological Bulletin*, 1974, *81*, 207–217.

DENNENBERG, V. H., AND MORTON, J. R. C. Effects of preweaning and postweaning manipulation upon problem-solving behavior. *Journal of Comparative and Physiological Psychology*, 1962, *55*, 1096–1098.

DUKES, W. F. "$N = 1$." *Psychological Bulletin*, 1965, *64*, 74–79.

EBBINGHAUS, H. *Memory.* (H. A. Ruger and C. E. Bussenius, Trans.) New York: Dover Publications, 1964. (Originally published 1885.)

EDWARDS, W., LINDMAN, H., AND SAVAGE, L. J. Bayesian statistical inference for psychological research. *Psychological Review*, 1963, *70*, 193–242.

FLEENER, D. E., AND CAIRNS, R. B. Attachment behaviors in human infants: Discriminative vocalization on maternal separation. *Developmental Psychology*, 1970, *2*, 215–223.

GAFFAN, D. Recognition impaired and association intact in the memory of monkeys after transection of the fornix. *Journal of Comparative and Physiological Psychology*, 1974, *86*, 1100–1109.

GALTON, F. *English Men of Science: Their Nature and Nurture.* (2nd ed.) Forest Grove, Ore.: Cass, 1970. (Originally published 1874.)

GATEWOOD, R. D., AND PERLOFF, R. An experimental investigation of three methods of providing weight and price information to customers. *Journal of Applied Psychology*, 1973, *57*, 81–85.

GOLDING, W. G. *Lord of the Flies.* New York: Coward, McCann & Geoghegan, 1954.

GOTTMAN, J. M. *N*-of-1 and *N*-of-2 research in psychotherapy. *Psychological Bulletin*, 1973, *80*, 93–105.

GOTTSDANKER, R., AND WAY, T. C. Varied and constant intersignal intervals in psychological refractoriness. *Journal of Experimental Psychology*, 1966, *72*, 792–804.

Grocers moan, but New York moves on unit prices. *Advertising Age*, 1969, *40*, 3.

GUILFORD, J. P., AND FRUCHTER, B. *Fundamental Statistics in Psychology and Education.* (5th ed.) New York: McGraw-Hill, 1973.

HARDIN, G. *Exploring New Ethics for Survival: The Voyage of the Spaceship Beagle.* New York: Viking, 1972.

HARPER, R. S., AND STEVENS, S. S. A psychological scale of weight and a formula for its derivation. *American Journal of Psychology*, 1948, *61*, 342–351.

HAYS, W. L. *Statistics for the Social Sciences.* (2nd ed.) New York: Holt, Rinehart and Winston, 1973.

HICK, W. E. On the rate of gain of information. *Quarterly Journal of Experimental Psychology*, 1952, *4*, 11–26.

HOLTZMAN, W. Statistical models for the study of change in a single case. In C. Harris (Ed.), *Problems in Measuring Change.* Madison: University of Wisconsin Press, 1963.

HYMAN, R. Stimulus information as a determinant of reaction time. *Journal of Experimental Psychology*, 1953, *45*, 188–196.

JACOBS, G. H., AND ANDERSON, D. H. Color vision and visual sensitivity in the California ground squirrel (*Citellus beecheyi*). *Vision Research*, 1972, *12*, 1995–2004.

JENSEN, G. D. Preference for bar pressing over "freeloading" as a function of number of rewarded presses. *Journal of Experimental Psychology*, 1963, *65*, 451–454.

KENNEDY, J. E., AND LANDESMAN, J. Series effects in motor performance studies. *Journal of Applied Psychology*, 1963, *47*, 202–205.

KENNEDY, J. L. A methodological review of extra-sensory perception. *Psychological Bulletin*, 1939, *36*, 59–103.

KEPPEL, G. *Design and Analysis, A Researcher's Handbook.* Englewood Cliffs, N.J.: Prentice-Hall, 1973.

KLUGH, H. E. *Statistics, the Essentials for Research.* New York: Wiley, 1970.

KOFFER, K., AND COULSON, G. Feline indolence: Cats prefer free to response-produced food. *Psychonomic Science*, 1971, *24*, 41–42.

KRAFT, C. L., AND ELWORTH, C. L. Night visual approaches. *Boeing Airliner*, 1969 (March-April), 2–4.

LA VOIE, J. Punishment and adolescent self-control. *Developmental Psychology*, 1973, *8*, 16–24.

LEWIN, K., LIPPITT, R., AND WHITE, R. K. Patterns of aggressive behavior in experimentally created "social climates." *Journal of Social Psychology*, 1939, *10*, 271–299.

LOCKARD, R. B. The albino rat: A defensible choice or a bad habit? *American Psychologist*, 1968, *23*, 734–742.

NEURINGER, A. J. Animals respond for food in the presence of free food. *Science*, 1969, *166*, 399–401.

ORWELL, G. *Nineteen Eighty-Four: A Novel.* New York: Harcourt Brace Jovanovich, 1949.

POULTON, E. C. Unwanted range effects from using within-subject experimental designs. *Psychological Bulletin*, 1973, *80*, 113–121.

POULTON, E. C., AND FREEMAN, P. R. Unwanted asymmetrical transfer effects with balanced experimental designs. *Psychological Bulletin*, 1966, *66*, 1–8.

PRATHER, D. C. Prompted mental practice as a flight simulator. *Journal of Applied Psychology*, 1973, *57*, 353–355.

PREMACK, D., SCHAEFFER, R. W., AND HUNDT, A. Reinforcement of drinking by running: effect of fixed ratio and reinforcement time. *Journal for the Experimental Analysis of Behavior*, 1964, *7*, 91–96.

RAVEN, J. C. *Progressive Matrices.* London: Lewis, 1947.

ROETHLISBERGER, F. J., AND DICKSON, W. J. *Management and the Worker.* Cambridge, Mass.: Harvard University Press, 1946.

ROSENTHAL, R. *Experimenter Effects in Behavioral Research.* New York: Appleton-Century-Crofts, 1966.

RUNQUIST, W. N. Verbal behavior. In J. B. Sidowski (Ed.), *Experimental Methods and Instrumentation in Psychology.* New York: McGraw-Hill, 1966.

SIEGELMAN, E., BLOCK, J[ack], BLOCK, J[eanne], AND VON DER LIPPE, A. Antecedents of optimal psychological adjustment. *Journal of Consulting and Clinical Psychology*, 1970, *35*, 283–289.

SIMON, J. R., AND RUDELL, A. P. Auditory S-R compatibility: the effect of an irrelevant cue on information processing. *Journal of Applied Psychology*, 1967, *51*, 300–304.

SINGH, D. Preference for bar pressing to obtain reward over freeloading in rats and children. *Journal of Comparative and Physiological Psychology*, 1970, *73*, 320–327.

SINGH, D., AND QUERY, W. T. Preference for work over "freeloading" in children. *Psychonomic Science*, 1971, *24*, 77–79.

STERNBERG, S. The discovery of processing stages: Extension of Donders' method. *Acta Psychologica*, 1969, *30*, 276–315.

TAYLOR, H. C., AND RUSSELL, J. T. The relationship of validity coefficients to the practical effectiveness of tests in selection. *Journal of Applied Psychology*, 1939, *23*, 565–578.

Textile industry. *Encyclopaedia Britannica, Macropaedia*, 1974, *18*, 170–189.

THORNDIKE, E. L. *Human Learning.* New York: Appleton-Century-Crofts, 1931.

THORNDIKE, E. L. *The Fundamentals of Learning.* New York: Bureau of Publications, Teachers College Press, 1932.

TORGERSON, W. S. *Theory and Methods of Scaling.* New York: Wiley, 1958.

WAGENAAR, W. A. Note on the construction of digram-balanced Latin squares. *Psychological Bulletin*, 1969, *72*, 384–386.

WALLIS, W. A., AND ROBERTS, H. V. *Statistics, a New Approach.* New York: Free Press, 1962.

WESTON, H. C., AND ADAMS, S. *The effects of noise on the performance of weavers.* Report No. 65, Part II. Industrial Health Research Board, London: Her Majesty's Stationery Office, 1932.

WHITE, R. W. Motivation reconsidered: The concept of competence. *Psychological Review*, 1959, *66*, 297–333.

WISE, R. A., AND DAWSON, V. Diazepam-induced eating and lever pressing for food in sated rats. *Journal of Comparative and Physiological Psychology*, 1974, *86*, 930–941.

YERKES, R. M. *The Dancing Mouse*. New York: Macmillan, 1907.

YERKES, R. M., AND DODSON, J. D. The relation of strength of stimulus to rapidity of habit formation. *Journal of Comparative Neurology and Psychology*, 1908, *18*, 459–482.

INDEX